HYDROGEN

HYDROGEN

The Essential Element

JOHN S. RIGDEN

HARVARD UNIVERSITY PRESS

Cambridge, Massachusetts
London, England
2002

Library of Congress Cataloging-in-Publication Data
Rigden, John S.
   Hydrogen : the essential element / John S. Rigden.
      p. cm.
   Includes bibliographical references and index.
   ISBN 0-674-00738-7 (alk. paper)
   1. Hydrogen.   2. Science—Methodology.   I. Title.
QD181.H1 R54 2002
546′.2—dc21   2001051708

MAY 23 2002

Contents

Prologue

Hydrogen is the most important constituent of the universe.
—Gerhard Herzberg

The heroine of this book is nature's simplest atom, the hydrogen atom. With one exception—the helium atom—hydrogen is the mother of all atoms and molecules. The hydrogen atom consists of a single electron and a single proton; the proton is the nucleus of the hydrogen atom and serves as the electron's anchor. The universe is teeming with hydrogen: every cubic centimeter of dark interstellar space, essentially void of any other known matter,[1] contains a few atoms of hydrogen. At the other extreme, every cubic centimeter of the planet Jupiter's interior contains in excess of 10 million billion billion (10^{25}) atoms of hydrogen. And every star, throughout its long life, illuminates its cosmic neighborhood with light that originates with the burning of the atom that dominates its material composition—hydrogen.

One must not dismiss this chemical element because of its simplicity. In fact, it is the simplicity of the hydrogen atom that has enabled scientists to unravel some of the mysteries of nature. This humble atom has consistently surprised the most distinguished (and confident) scientists and contributed to our understanding of the natural world.

This book, however, is more than a book about the hydrogen atom. It is a drama, written for the general reader, in which the intriguing hydrogen atom plays a starring role. Each chapter unfolds a particular episode in which hydrogen has led scientists to new scientific insights.

Collectively, the twenty-three chapters that follow reveal much about the conduct of science. On one level it is a focused story that chronicles the hold the simplest atom has had on the minds of the world's greatest scientists over the decades reaching back into the nineteenth century. Niels Bohr, Arnold Sommerfeld, Otto Stern, Werner Heisenberg, Wolfgang Pauli, Erwin Schrödinger, Paul Dirac, Harold Urey, I. I. Rabi, Norman Ramsey, Edward Purcell, Felix Bloch, Willis Lamb, Daniel Kleppner, and Theodor Hänsch all have advanced and refined knowledge of the physical world through their fascination with the hydrogen atom.

On another level, the story of hydrogen reveals how science is conducted. Physical theories are created to provide explanatory schemes whereby the observed world can be understood with quantitative precision. Those theories that capture the support of scientists are those that allow detailed predictions to be made and lead to new insights into the natural world. Good theories are simple theories that unite disparate realms of experience. Physical theories, however, must always yield to the demands of experimental data. Experimental facts are incontrovertible. If they are not accommodated by theory, the theory is held in question. Theories, good theories, are not quickly abandoned. Strenuous effort is exerted to refine a good theory so that experimental facts can be explained. In the final analysis, however, experimental results, once tested and retested, once verified by independent experimental methods, ultimately rule. Dirac's theory was elegant and beautiful, but in the face of data from Lamb and Rabi, it fell short. Their data then became the stimulus for the more powerful theory of quantum electrodynamics.

The experiments on the hydrogen atom chronicled in these chapters demonstrate the significance of precise measurements. Although all scientists seek to refine their experimental procedures to minimize the uncertainties in their measured results, uncertainties of several percent are typical. However, to expose shortcomings in theories and to test their limits, precise results are often necessary. The hydrogen atom has been the premier physical system for challenging theoretical constructs and precise measurements are the *sine qua non* when hydrogen is the subject of investigation. Furthermore, precise measurements can reveal unexpected results. In Rabi's series of experiments to measure the magnetic moments of the proton and deuteron, uncertainties were reduced from 26 percent to 0.7 percent. With the improved precision, evidence for a new property of the nucleus, the quadrupole moment, was found lurking in the data.

Through the example of hydrogen, we have also seen how basic science may lead to practical applications. Basic science typically operates far from the technological applications that predictably follow. The objective of basic science is to learn how the world works. Nonetheless, the knowledge gained through basic research and the methods developed to probe the natural world frequently hold within them the potential for very practical and welcome uses. The magnetic resonance method discovered by Rabi and his group of students led to nuclear magnetic resonance at the hands of Purcell and Bloch, which in turn led to the powerful medical diagnostic tool of magnetic resonance imaging. Ramsey's and Kleppner's hydrogen maser clock is an integral part of the technology of global positioning systems, which have manifold applications.

When nature's ways are understood, applications follow that can be used for good or bad, for peace or war. Consider the fusion of hydrogen. Einstein's relativity theory, basic physics at its best, showed how nuclear fusion could produce vast amounts of energy. Applications were soon understood. On the one hand, for example, it was understood that the fusion of hydrogen occurs in the Sun and its energy nurtures life on planet Earth. On the other hand, the fusion of hydrogen can occur in a bomb and its energy can inflict devastating destruction. The hydrogen bomb is an important part of the hydrogen story and it could have been the subject of a chapter in this book. I decided against it for two reasons. First, the prominent theme of the following chapters is how the hydrogen atom led to new basic scientific knowledge. The fusion bomb does not fit into that theme. Second, there is a vast literature on the hydrogen bomb and another chapter seemed hardly necessary.

Science is an international enterprise, which the examples in this book make clear. Although communities may differ enormously in their cultures, their religious convictions, their artistic expressions, and their political structures, in the arena of science, the world's diverse human groups are unified. There is no German science, no Asian science, no Hindu science. Bose was Indian, Einstein was German, but the two came together as scientists and predicted a new form of matter—the Bose-Einstein condensate, which was eventually verified by American scientists.

This book further illustrates how science itself has changed over the decades. The early chapters typically have one name associated with them. In earlier eras, science was such that an individual could work alone and make significant contributions. The experimental apparatus was relatively simple, could be constructed by one scientist, and put together on a laboratory table. As science progressed through the twentieth century, however, it became more specialized, and the experimental apparatus required became more complex. Many talents are now required to conduct an experiment and science has become a group activity. Many scientists have measured the Rydberg constant and could have been identified along with Hänsch. Four experimental physicists were identified with the discovery of the Bose-Einstein condensate, and no one was identified with the discovery of antihydrogen simply because many scientists at different laboratories were involved.

The hydrogen atom has intrigued physicists because its simplicity allows conceptual models to be created and then tested against experimental data. The inherent logic of a conceptual model is expressed mathematically and the simplicity of the hydrogen atom permits the resulting mathematical expressions to be solved exactly and compared directly with experimental data. This is physics at its best.

At various times in the history of physics, there has been a tendency for physicists to believe that the time to unravel the final mysteries of nature was at hand. In response to this malady, I once wrote a short piece entitled "H Stands for Hydrogen . . . and Humility."[2] (This piece, I am told, hung for a period on an office wall at CERN, the high energy physics laboratory in Geneva, Switzerland.) In the essay I raised a cautionary note about claims that we were nearing a "grand unified theory" that would explain all physical interactions or that we were nearing a complete understanding of such momentous questions as how the universe began. "The hydrogen atom," I wrote, "still beckons."

1

In the Beginning:
Hydrogen and the Big Bang

If God did create the world by a word, the word would have been hydrogen.
—Harlow Shapley

The story of hydrogen begins before there was anyone to notice. Long before the Earth and its planetary siblings existed, before the Sun and the Milky Way existed, and even before chemical elements like oxygen, sodium, iron, and gold existed, the hydrogen atom was old, old news.

According to current wisdom, our universe began about 15 billion years ago at a point with infinite density and infinite temperature. That was the beginning of time; that was the origin of space. Since then, the original point has expanded in all directions to the dimensions of the current universe. As the universe expanded, the cosmic clock ticked and the temperature cooled: at 0.01 second after the big bang, the temperature was 100,000 billion degrees; 0.11 second, 30,000 billion degrees; 1.09 seconds, 10,000 billion degrees; 13.82 seconds, 3,000 billion degrees. By the time the universe was four minutes old, the basic ingredients required for all that was to follow were present and their basic modes of interaction were established. The stage was set for everything that followed.[1]

Hydrogen is the simplest of all atoms. In its dominant form, hydrogen consists of one electron and one proton; in its rare form, called deuterium, there are three particles: an electron, proton, and a neutron. By contrast, ordinary water, a simple molecule, consists of twenty-eight particles: ten electrons, ten protons, and eight neutrons. The water molecule is very complicated when compared to the hydrogen or deuterium atoms. Because of its simplicity, hydrogen dominates the 15 billion-year tale of our universe. Approximately 300,000 years after the origin of our universe, the temperature had cooled to approximately 3,000 degrees and the hydrogen and helium atoms took their characteristic forms. Even this early, a particular kind of universe was inevitable: a universe that would eventually become a hospitable haven for life.

When atoms first began to take form, the ingredients available were limited. There were photons (particles of light) and neutrinos, and elementary particles of matter—electrons and protons (the nucleus of the hydrogen atom is a proton). There were composites of elementary particles—deuterons, a proton plus a neutron (the deuteron is a special part of the story told in this book because it is the nucleus of the heavy hydrogen atom, deuterium), and alpha particles, two protons plus two neutrons (the nucleus of the helium atom is an alpha particle). By the time the universe was 300,000 years old, neutrinos were aloof from their surroundings and did not participate in the birth of atoms, and photons were not essential to the atom-forming process. So, to form the first atoms of our universe there were electrons, protons, deuterons, and alpha particles. In this mix, protons outnumbered alpha particles by about eleven to one. The deuteron was a mere sprinkling in the mix. Thus, when atoms formed, the ingredients present coupled with the particle recipes for hydrogen and helium resulted in an atomic mix of about 92 percent hydrogen, 8 percent helium, and a fraction of a percent deuterium. Today, 15 billion years after hydrogen and helium were first formed, these elements remain the most abundant throughout the cosmos: hydrogen makes up approximately 90 percent of the total, whereas helium comes in at about 9 percent.

Since the ingredients for hydrogen and helium atoms—elec-

Figure 1.1 A cosmic cloud of hydrogen, where stars are born, in the form of a pillar, as seen by the Hubble Space Telescope. The globules are forming stars. This picture of this cloud, in M16, was taken by John Hester and P. Scowen in 1995.

trons, protons, and neutrons—were present in the earliest seconds of the universe, why did it take 300,000 years before atoms appeared? Dropping temperatures over this span of years slowed the rapidly moving protons and electrons to speeds that allowed the electrical attraction between them to challenge their independent motions, bring them together, and form stable atoms. In fact, even the strongest force of nature, the nuclear force, was not strong enough to pull the frantic protons and neutrons together into nuclei during the earliest seconds of the universe. It was not until the universe was about fourteen seconds old and had expanded and cooled considerably that the first nuclei, alpha particles, formed. The early formation of alpha particles testifies to their stability. Deuterons, while simpler than alpha particles, are not as stable. Consequently, they did not form until the universe was almost four minutes old.

The primordial period of nuclear synthesis was all over by the time the universe was four minutes old. Nuclei heavier than that of helium—nuclei of beryllium, boron, and carbon, for example—did not form because these heavier nuclei could not compete with the inherent stability of the helium nucleus. Thus, all the free neutrons that were still available at the four-minute point took refuge in either the helium nucleus or the heavy hydrogen nucleus.

Essentially all the heavy hydrogen in the universe today originated during the first minutes of cosmic time. One thousand tons of heavy water, used to detect solar neutrinos, fill the tank at the Sudbury Neutrino Observatory in Sudbury, Ontario. This heavy water, each molecule of which consists of one oxygen atom, one hydrogen atom, and one deuterium atom, brings together deuterium that was formed when the universe was about four minutes old. When you hold a tube of heavy water in your hand, you hold primordial atoms, remnants from the first moments after the big bang.

Today, 410 million billion seconds after the big bang, the temperature of the universe has dropped to three degrees above absolute zero. Embedded in this frigid environment are galactic systems distributed across the far reaches of the observable universe. Each galaxy consists of stars and dust clouds. Each star, each dust cloud in each and every galaxy consists of about 90 percent hydrogen atoms and 9 percent helium atoms. Because of this composition, established approximately 15 billion years (or 410 million billion seconds) ago, the stars twinkle and the Sun shines.

The Sun is a typical star. The composition of the Sun (as well as other stars) reflects the cosmic abundance: about 90 percent of the atoms making up the Sun are hydrogen. And it is the fusion of hydrogen that fuels the Sun. Every second, 600 million tons of hydrogen are fused into helium in the core of the Sun, releasing prodigious energy that slowly makes its way from the core to the Sun's surface, heating it to a temperature of 5,800 K. The Earth, 92 million miles away, basks in this life-giving warmth.

Approximately 3.5 billion years ago, life emerged on at least one planet orbiting one star. There may be planets other than Earth that nurture life: we simply do not know. On planet Earth, hydrogen remained obscure for many centuries. Paracelsus (born Theophrastus Bombast von Hohenheim) noted during the early years of the sixteenth century that when acids attacked metals, flammable gas was a by-product. He had unknowingly observed hydrogen. Other chemists and physicists produced hydrogen and in 1671 Robert Boyle described its properties. As is frequently the case in science, the credit for discovering hydrogen rests on how "discovery" is defined. The credit for isolating and characterizing hydrogen goes to Henry Cavendish, who isolated hydrogen and determined its density in 1776. The French chemist Antoine-Laurent Lavoisier, whose head was severed by the guillotine on May 8, 1794, gave hydrogen its name.

The world as we know it is a consequence of the balance between the number of hydrogen nuclei and the number of helium nuclei, established in the early moments after the big bang. Perhaps it is preferable to say that the world is a consequence of the basic laws that *produced* this particular blend of hydrogen and helium. Did the laws of nature exist prior to the origin of the universe? Did the laws of nature take their present form at the instant of the big bang? One millionth of a second after the big bang? No one can say. Looking back, however, we can say the following: if the weak force had been just a little weaker, the free neutron would decay a little more slowly and, as a result, the universe would have started out as predominantly helium rather than hydrogen. A world without hydrogen is a world without water, a world without carbohydrates, a world without proteins—a world without life.

So take your pick. We can say that the world is the way it is because the laws of nature are the way they are. Or we can say that the world is the way it is because hydrogen is the way it is. Whichever you select, one or the other, is a matter of preference. Either way, the little hydrogen atom commands the stage on which the long and enchanting drama of our universe, the story of galaxies, stars, planets, and life, unfolds.

2

Hydrogen and the Unity of Matter: The Prout Hypothesis

William Prout, 1815

Hydrogen seems almost aware of its illustrious history, for the atom behaves in a regal fashion.

—Daniel Kleppner

All of us, from infant to most senior, are aware of the world around us. The breadth and depth of this awareness varies dramatically from person to person, but the same world is there for all to observe. When we look not only with our eyes, but also with our minds and souls, we can see very different worlds.

For example, we might interpret the world as diverse: each face in a throng of people is unique; each letter of the alphabet is explicitly different; each planet in the solar system has a distinct size and a definite orbital location in the Sun's family of satellites. On the other hand, we might look at the same world and see unity: the eyes, nose, and mouth together possess the corporal unity that is the human face; the alphabetical letters in proper sequence have a functional unity as a language; and the planets—regardless of size and composition, regardless of distance from the Sun—exhibit a spherical shape and travel along a circum-solar elliptical path that manifests the unity of physical law.

Diversity or unity? The answer depends on the phenomenon of perception and the magic of mind. For Dr. William Prout (1785–

1850), the answer was unity and that unity, he claimed, was based on the hydrogen atom.

Prout's idea of an underlying unity for all matter originated during his student days. The influences that prompted this mindset cannot be identified with certainty. Was its origin religious? Prout, a deeply religious man, wrote the treatise *On the Power, Wisdom and Goodness of God, as Manifested in the Creation*. Was it music? Prout was an accomplished organist. Was it medicine? Prout was a physician who specialized in urinary and digestive disorders. Was it the work of other scientists? He was especially influenced by the works of British chemists Sir Humphry Davy (1778–1829) and John Dalton (1766–1844). Whatever the origin of his conviction, Prout hoped to develop a more analytical, unified chemistry. He thereby joined scores of other scientists whose approach to their science was strongly influenced by deeply held convictions about the correct way to seek an understanding of natural phenomena.

Prout believed that a primordial substance—some basic stuff— lies under the diversity of the material objects comprising the universe, and that this basic stuff is hydrogen. The idea of a primordial substance was not new. Thales, who lived on the isle of Miletus some 2,400 years earlier, had concluded that water was the basis of the manifold forms of all material objects. Never mind that he was wrong. Thales' idea was extremely provocative. The original Greek atomists, Leucippus and Democritus, who lived and thought about the world a century after Thales, continued the same intellectual quest to identify the underlying unity of matter. Through the centuries and to the present day, the quest continues. Prout was a part of that tradition.

By the time Prout received his medical degree from Edinburgh University in June 1811, chemists were flirting with the idea of atoms. No one had seen an atom, no one knew the nature of an atom, and there were reputable scientists who rejected the idea of atoms altogether. Chemists did recognize, however, that certain substances possessed established properties that defined their identity. These substances were the chemical elements. Whether or not the basis for their distinct identity was atoms, however, remained unresolved. During the early years of the nineteenth century, Dalton transformed atomistic philosophizing into atomic experimentation and thereby amassed evidence that eventually provided strong support for a fully credible atomic theory of matter.

Prout, a contemporary of Dalton, was one of these early experimental chemists. At that time there was no periodic chart of the chemical elements and many of the elements had yet to be discovered. However, some forty elements were known—a sufficient number to prompt a man of Prout's passion to seek unity in this diversity of chemical elements. In the tables he developed to organize his results, Prout applied his ideas to forty-two elements.

The atoms of each chemical element are characterized by a definite weight: the atoms of any one element all have the same weight, but the atoms of some elements are heavier than those of others. Oxygen atoms, for example, weigh more than those of nitrogen. The hydrogen atom, the simplest of all atoms, weighs the least. If, Prout reasoned, hydrogen was the fundamental building block of all the heavier atoms, then the atomic weights of all elements should be exact multiples of the atomic weight of hydrogen. This is what Prout set out to prove.

Prout did not consider himself a proficient experimentalist; nonetheless, he designed and carried out experiments to determine the weights of such atoms as iodine, phosphorus, sodium, iron, zinc, potassium, and beryllium. For other elements he accepted the atomic weights that had been measured by scientists he considered trustworthy. Of critical importance was the atomic weight accepted for hydrogen itself and for this Prout used the value measured by Davy. With these data in hand, Prout proceeded to show in a table that the weights of the heavier elements were exact multiples of the weight of hydrogen. For example, the weight of carbon is six times the weight of hydrogen, the weight of nitrogen is fourteen times the weight of hydrogen, the weight of potassium is forty times the weight of hydrogen, and the weight of iodine is 124 times the weight of hydrogen. And so it went with the other elements Prout examined. For him the results were convincing. "Others [chemical elements] might doubtless be mentioned," concluded Prout in his 1815 paper, "but I submit the matter for the present to the consideration of the chemical world." In 1816 Prout anonymously submitted his paper and a follow-up paper each with the same title, "On the Relation Between Specific Gravities of Bodies in the Gaseous State and the Weights of Their Atoms."[1] Soon after, Prout identified himself as the author of these two papers and his idea became known as Prout's hypothesis.

Prout delivered his hypothesis into a scientific world whose practitioners held opposing views as to how one should approach the study of the chemical elements. On the one hand, there were those who thought that chemists should focus entirely on the facts that came out of careful experimentation and avoid hypothetical speculations about the deeper nature of matter. Representing this point of view were such first-rate chemists as Dalton and the Swedish chemist J. J. Berzelius (1779–1848). On the other hand, there were those for whom the lure of a generalization that could unite the elements into a coherent theory of matter was enticing. Physicists generally fell into this latter group along with some physically minded chemists. Michael Faraday, for example, said to William Crookes, "To discover a new element is a very fine thing, but if you could decompose an element and tell us what it is made of—that would be a discovery indeed worth making."[2] Many scientists of the nineteenth century, like William Herschel and James Clerk Maxwell, believed that "atoms bear the impress of manufactured articles."[3] The question to be answered was, "What is the raw material from which these 'manufactured' atoms are made?" Prout believed it was hydrogen.

Regardless of the predisposition of the scientist, Prout's hypothesis initiated decades of careful research, by proponents and opponents alike, designed to test the validity of Prout's idea. Experiments were conceived to measure with the greatest possible accuracy the atomic weights of the chemical elements. As new elements were discovered, they were put to the Proutian test. Through the decades leading up to the First World War, Prout's hypothesis was neither proved nor disproved. In 1886, sixty years after Prout's hypothesis was published, Crookes delivered his presidential address before the chemistry section of the British Association for the Advancement of Science. In this address he acknowledged the differences between Prout's hypothesis and the known atomic weights. "Still," he said, "in no small number of cases the actual atomic weights approach so closely to those which the hypothesis demands that we can scarcely regard the coincidence as accidental."[4]

The first hint of the actual raw material out of which atoms are constructed came with the discovery of the electron by J. J. Thomson in 1897. Then Ernest Rutherford, fourteen years later, discovered the atomic nucleus. After Rutherford's 1911 finding, the idea that atoms were made of negatively charged electrons and a positively charged massive core quickly gained widespread acceptance, and further discoveries followed rapidly. But first, we must back up a few years to pick up another strand of history.

In 1896, the year before Thomson discovered the electron, Antoine Henri Becquerel discovered radioactivity. With this finding the long-assumed immutability of atoms became untenable. For the next decade and more, many physicists analyzed the decay products of atoms and by 1910 found that the decay products of radioactive atoms involved "daughter" atoms that were sometimes chemically identical to the "mother" atom, but differed in their atomic weight. Such atoms—identical in their chemical behavior, but different in their inherent physical character—were called isotopes in Frederick Soddy's 1913 paper in *Nature*.[5] Were isotopes a consequence of the radioactive process or were isotopes more general? In other words, did nonradioactive atoms also come in different isotopic species? The Great War delayed the answer to this final question, but during the war, Thomson conceived of a new instrument that would, he thought, allow the masses of atoms to be measured with unprecedented accuracy. After the war, Thomson's colleague Francis William Aston built the first "accurate" mass spectrograph and showed in 1920 that the stable element neon had two isotopes with atomic masses of twenty and twenty-two. As additional elements were analyzed by Aston's technique, it was established that the atomic weights of heavy elements were not exactly whole-number multiples of the atomic weight of hydrogen.

Prout's intriguing hypothesis was proven false by the accumulation of incontrovertible evidence gathered at the laboratory bench. Hydrogen, the heroine of this book, was not destined to be the fundamental building block of all other atoms. But in 1815 the simplicity of hydrogen made it the most obvious candidate for an empirical theory of matter; thus, hydrogen stimulated an idea that transcends the details that gave the idea expression. The question articulating the idea is eternally seductive: Is there some irreducible basic stuff at the foundation of the material world? Thales' proposal of water was wrong; Heraclitus' answer, fire, was also wrong. The four elements of the ancient world—earth, water, air, and fire—proved inadequate. Prout's hydrogen, after decades of accumulated evidence, could not be supported as the answer. The electron and the proton, for a while, seemed to provide the holy grail of matter; but under close examination, protons lost this privileged status.

What is the basic stuff? Is the answer electrons plus the quarks and gluons that make up the proton and neutron? No one knows with absolute certainty, but the quest continues. And the quest is driven by the same powerful urge that compelled Thales and Prout to see unity in the diversity of the material world. It was the same urge that brought Isaac Newton to see unity in movement on Earth and those quiet motions we observe in the night's sky; it was the same urge that brought Maxwell to see the diverse behaviors of electricity, magnetism, and light exhibiting such a unity. In his synthesis, Maxwell captured this unity. Contemporary physicists seek a similar sort of unity among the four basic forces that they believe account for all the physical pageantries we witness in the observable universe.

The quest for unity is a staple of physics. In the chapters that follow, we shall see that the simple hydrogen atom has been a vital presence in that quest.

3

Hydrogen and the Spectra of the Chemical Elements: A Swiss High School Teacher Finds a Pattern

Johann Jakob Balmer, 1885

Historically, the simple and regular Balmer spectrum has inspired . . . pathbreaking discoveries.

—Theodor W. Hänsch

From 1859 until his death at age seventy-three, Johann Jakob Balmer (1825–1898) was a high-school teacher at a girls' school in Basel, Switzerland. His primary academic interest was geometry, but in the mid-1880s he became fascinated with four numbers: 6,562.10, 4,860.74, 4,340.1, and 4,101.2. These are not pretty numbers, but for the mathematician Balmer, they became an intriguing puzzle: Was there a pattern to the four numbers that could be represented mathematically? The specific numbers that commanded Balmer's attention were four of many, many such numbers Balmer could have examined. But the four numbers Balmer chose were special because these numbers pertained to the atom of hydrogen. We shall return to these numbers shortly.

The significance of an everyday object often reaches far beyond its own apparent simplicity. A little toy compass whose pivoting pointer mysteriously orients itself along a north-south direction was a source of inspiration to the young Albert Einstein—the sense of awe it inspired in him never waned. A glass prism captures the bright light of the Sun or the feeble glimmer of a candle and sparkles with surprising brilliance. With such a simple glass prism, Isaac Newton demonstrated that the Sun's white light was not what it seemed: it was, instead, a mixture of many pure colors.

Most of what we know about the material makeup of the universe, from the Sun that commands our solar system to the minerals that make up the Earth's crust, has come by examining in detail how atoms either absorb or emit light. In order to learn about the properties of atoms, however, a way must be found to examine individual wavelengths of light. Since the light from most sources is, like sunlight, a composite of many wavelengths, the challenge is to separate the composite into its individual wavelength parts. This is what the glass prism achieves for sunlight. Take a glass prism from a chandelier and you hold in your hands the means to probe into the atomic nature of matter.

When a narrow beam of sunlight enters one side of a prism, the beam bends slightly and then emerges from the prism as a broadened beam displaying the colors of the rainbow: red, orange, yellow, green, blue, indigo, and violet. The different wavelengths associated with these colors—from the longer wavelength of red light to the shorter wavelength of violet light—make up the visible spectrum. However, these colors constitute only a small part of the radiant energy coming from the Sun. In 1800, William Herschel (1738–1822), who discovered the planet Uranus, used a thermometer to determine the heating effects of light with different colors. He found that the temperature increased as he moved the thermometer away from the violet toward the red light, but, more interestingly, that the heating effect continued to increase as he moved the thermometer out of the red light into the darkened region beyond the red. From this, he correctly inferred the presence of invisible light, which we now call the infra-red region of the spectrum. In 1801, the German physicist Johann Wilhelm

Ritter (1776–1810) discovered the presence of another invisible radiation at the other end of the visible spectrum, beyond the violet or, as we now call it, the ultraviolet.

From the beginning of the nineteenth century, scientists relied on the glass prism as an active element in optical experimentation. During the year following the discoveries of Herschel and Ritter, the British scientist William Wollaston (1766–1828) made a seminal discovery in a way that established the terminology scientists still use today. Up to this time, scientists had followed Newton's example, allowing sunlight to pass through a small circular hole in an opaque shield. Through the hole came a beam of sunlight with a cross section like the hole itself: circular. Wollaston changed this. He cut a slit in the barrier, and from this slit, a ribbon of light fell on his glass prism. When he examined the Sun's visible spectrum, he noticed several dark images of the slit. Wollaston concluded that the dark images represented certain wavelengths in the visible light coming from the Sun that were missing, and these missing wavelengths revealed themselves as missing light, or dark lines in the spectrum. The dark images in the solar spectrum came to be called spectral lines.

The dark lines discovered by Wollaston quickly attracted the attention of other scientists. Joseph Fraunhofer (1787–1826) observed 574 dark lines in the solar spectrum and he labeled and mapped the more prominent ones. Further, and most significantly, Fraunhofer found that the two *dark* lines in the solar spectrum, which he labeled "D," coincided in position with the two *bright* lines from the sodium lamp he had in his laboratory. Fraunhofer did not explicitly link these two observations, but this coincidence between the light from the Sun and that from a light source on Earth was a coincidence that awaited further explanation. Fraunhofer did more: he examined the light from the planets and found patterns of spectral lines similar to those he had observed in the Sun's light. He also examined the light from Sirius and other bright stars and he found both consistencies and differences in the spectral line patterns from one star to another.

By this time, scientists were studying light from as many sources as they could conjure. In 1822, the Scotsman David Brewster (1781–1868) invented a device that, by means of a flame, vaporized small amounts of material. The light from this vaporized material could then be studied. He added 1,600 new spectral lines to those discovered by Fraunhofer and other investigators. During the same year, 1822, John Herschel (1792–1871), William Herschel's son, vaporized various metallic salts and established that the light from the flames could be used to detect the presence of these metals in very small samples. A few years later, William Talbot (1800–1877) showed that the spectrum of each of the chemical elements was unique and that it was possible to identify the chemical elements from their spectra.

It often takes time for the implications of experimental data to be understood and to be acted upon. Fraunhofer's earlier observation that the solar D-lines coincided with the spectral lines of a sodium lamp eventually prompted further important experiments. In 1849, Jean Bernard Léon Foucault (1819–1868), a Parisian physicist, made an unexpected discovery. He passed sunlight through a vapor of sodium and he found that the solar D-lines were darker. His conclusion was that the sodium vapor "presents us with a medium which emits the rays D on its own account, and which absorbs them when they come from another quarter."[1] The consequences of Foucault's experiment, however, were expressed more cogently by Sir William Thomson (later Lord Kelvin). He drew the following explicit conclusion: "That the double line D, whether bright or dark, is due to the vapor of sodium . . . That Fraunhofer's double dark line D, of solar and stellar spectra, is due to the presence of vapor of sodium in atmospheres surrounding the Sun and those stars in whose spectra it has been observed."[2]

Thomson's recognition that the dark D-lines of the Sun's light were somehow connected with the bright lines of sodium light and that both were due to the element sodium can be cited as the beginning of astrophysics. But the foundation of spectroscopy was put in place in 1859 by Gustav Robert Kirchhoff (1824–1887) and Robert Bunsen (1811–1899). Kirchhoff repeated Fraunhofer's earlier experiment (without knowing that Fraunhofer had already done it) of passing sunlight through sodium vapor. Like Fraunhofer, he saw that the dark lines of the solar spectrum got darker when the Sun's light was passed through a vapor of sodium. Kirchhoff and Bunsen, however, articulated the general principle on which spectroscopy rests; namely, that under the same physical conditions, the emission of light by an element (which gives rise to the bright lines) and the absorption of light by the same element (which gives rise to the dark lines) produce spectral lines with identical wavelengths.

The vast array of numbers, thousands of numbers, representing the wavelengths of these spectral lines required an explanation. Was there an underlying pattern? If so, what was happening inside the atom to cause the observed pattern of spectral lines? George Johnstone Stoney (1826–1911) proposed in a 1868 paper that spectral lines were caused by some kind of periodic motion inside the atom. Arthur Schuster (1851–1934) refuted Stoney's idea in 1881, but concluded, "Most probably some law hitherto undiscovered exists."[3]

This brings us back to Balmer, the high-school mathematics teacher. By the time Balmer became interested in the problem, the spectra of many chemical elements had been studied and it was clear that each element gave rise to a unique set of spectral lines. Balmer was a devoted Pythagorean: he believed that simple numbers lay behind the mysteries of nature. Thus, his interest was not directed toward spectra *per se*, which he knew little about, nor was it directed toward the discovery of some hidden physical mechanism inside the atom that would explain the observed spec-

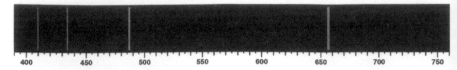

Figure 3.1 The visible spectrum of hydrogen, called the Balmer series.
The wavelengths of these spectral lines are, from left to right, 4,101.2 Å,
4,340.1 Å, 4,860.74 Å, and 6,562.10 Å.

tra; Balmer was intrigued by the numbers themselves. Was there
a pattern to the numbers? In the mid-1880s, Balmer began his
examination of the four numbers associated with the hydrogen
spectrum. At his disposal were four numbers measured by Anders
Jonas Ångström (1814–1874): 6,562.10, 4,860.74, 4,340.1, and
4,101.2. These numbers represent the wavelengths, in units of
angstroms, of the four visible spectral lines in the hydrogen spectrum (Figure 3.1).[4]

No one knows how many unsuccessful formulations Balmer attempted. What we *do* know is that in 1885 Balmer published a paper in which his successful formulation was communicated to the
scientific world. In this paper, Balmer showed that the four wavelengths could be obtained with the formula

$$\lambda = b \, \frac{m^2}{m^2 - n^2}.$$

In this formula, the wavelength λ is given in angstroms (Å). The
symbol b, which Balmer called "the fundamental number of hydrogen," has the numerical value of 3,645.6 Å; the symbol n is an
integer, which Balmer gave the value 2. The symbol m is another
integer, to which Balmer assigned the values starting with $m = 3$
and continuing with $m = 4$, 5, and 6. With $m = 3$, Balmer calculated one wavelength. With $m = 4$, another wavelength, and so
on. The result of Balmer's calculation was stunning:

Value of m	Balmer's calculated wavelengths	Ångstrom's measured wavelengths
m = 3	6,562.08 Å	6,562.10 Å
m = 4	4,860.80 Å	4,860.74 Å
m = 5	4,340.00 Å	4,340.10 Å
m = 6	4,101.30 Å	4,101.20 Å

A comparison of the wavelengths calculated by Balmer's formula with those measured by Ångström reveals their close agreement. Balmer had achieved his objective. He had found a mathematical formula that "expresses a law by which their wavelengths [hydrogen's] can be represented with striking precision."[5] But Balmer did more for science than simply develop a formula that reproduced the numbers representing the wavelengths of the four visible spectral lines of hydrogen. He suggested that there might be additional lines in the hydrogen spectrum. Specifically, Balmer extended his calculation by using the next integer, $m = 7$, and calculated a wavelength equal to 3,969.65 Å. As far as Balmer knew, this spectral line did not exist; so he was essentially making a prediction. What Balmer did not know was that Ångström had in fact already measured the wavelength of another spectral line with the value of 3,968.10 Å. Still other spectral lines with their own wavelengths were predicted by Balmer and later found by other scientists.

Ångström measured the wavelengths of the spectral lines of hydrogen, but Balmer showed that the wavelengths of these spectral lines are not arbitrary; rather, the value of the wavelengths are the expression of one particular mathematical formula. Balmer's work illustrates the hierarchy of values for physicists: discovering an underlying order in measured numbers often counts for more than the measurements themselves.

Balmer's formula had a striking effect on the scientific investigations of atomic spectra. To begin, it altered scientists' thinking about spectral lines. Before Balmer published his results, scientists drew an analogy between spectral lines and musical harmonics. They assumed that there were simple harmonic ratios between the frequencies of spectral lines. After Balmer's work, all scientists came to recognize that spectral wavelengths could be represented by simple numerical relationships. Even more, Balmer's success inspired scientists to believe that order lay beneath the confusing profusion of spectral lines.

In the closing paragraph of his paper, Balmer noted the "great difficulties" in finding the "fundamental number" of other chemical elements. He specifically mentioned the elements oxygen and carbon. Had Balmer chosen to apply his effort to any chemical element other than hydrogen, we would never have heard of the high-school teacher from Basel. He owed his success to a judicious choice: to study the spectral lines of hydrogen, the simplest chemical element. Through Balmer's success, the hydrogen atom prepared the way not only for an eventual understanding of atomic spectra, but also to an understanding of how spectral lines originate within the unseen atom.

Balmer unwittingly introduced a ticking bomb into the literature of physics—a bomb that would remain undisturbed for twenty-eight years. After he discovered his mathematical expression, Balmer disappeared from the ranks of working scientists and continued his classroom work teaching young ladies mathematics. Neither he nor his students recognized that his paper on the spectrum of hydrogen would bring him scientific immortality: the spectral lines of hydrogen that were the focus of Balmer's attention are now known as the Balmer series.

4

The Bohr Model of Hydrogen:
A Paradigm for the Structure of Atoms

Niels Bohr, 1913

[The Bohr model] scored a stunning success in accounting for major features
of the observed spectrum of the hydrogen atom.

—Bretislav Friedrich and Dudley Herschbach

"As soon as I saw Balmer's formula, the whole thing was immediately clear to me." How logically neat it would be if Balmer's simple formula had, out of the blue, led Bohr directly to his model of the hydrogen atom. Unfortunately, neat logic must give way to the reality of events as they occurred. Bohr *did* see Balmer's formula, he *did* make the statement quoted above, and he *did* proceed to develop his model of the hydrogen atom quickly. When Balmer's formula came to Bohr's attention, however, he was already deeply engaged in an intellectual struggle to develop a model of the hydrogen atom that, among other things, explained its spectral behavior. Bohr's recognition of the significance of Balmer's formula is a classic example of the prepared mind.

Niels Henrik David Bohr was born in Copenhagen, Denmark on October 7, 1885. Christian Bohr, his father, was a professor of physiology at the University of Copenhagen and his mother, Ellen Adler, came from a prominent Jewish family. Niels had one older sister, Jenny, and one younger brother, Harald. The family home was a place where Professor Bohr and his university colleagues gathered and young Niels was exposed to the ideas that animated intellectual discussions during the concluding years of the nineteenth century. The time of his birth was auspicious for a budding physicist: when Niels Bohr received his doctorate in 1911, the world of physics was pregnant with potential.

One of those whose discoveries prepared the field for Bohr and others was J. J. Thomson (1856–1940), who in 1884, at the age of twenty-eight, became Cavendish Professor of Experimental Physics at the University of Cambridge, following in the steps of James Clerk Maxwell (1831–1879) and Lord Rayleigh (1842–1919). In 1897, J. J. Thomson discovered "matter in a new state" and with this discovery it was clear, as Thomson wrote, "the subdivision of matter [had been] carried very much further."[1] Thomson's new state of matter eventually became the electron, and with its discovery most physicists understood that the atom had inner parts. Thomson's experimental measurements gave a single number that represented only the ratio of the electron's mass divided by its charge, m/e; thus Thomson could establish neither the electron's charge nor its mass separately. Thomson's data, however, provided two important clues. Hydrogen provided one of them. Thomson's result showed that the mass-to-charge ratio of hydrogen, as determined by electrolysis experiments, was 1,000 times larger than the same ratio for the electron. This allowed Thomson to conclude that the smallness of the electron's m/e ratio was due to either the smallness of m or the largeness of e (or a combination of the two). The second clue was that the ratio m/e had a negative value. Since mass is always positive, this meant that the charge carried by the electron was negative.

Before the discovery of the electron, many attempted to represent the atom by a model whose behavior would parallel the behavior of atoms. In his 1871 inaugural presidential address before the British Association for the Advancement of Science, Sir William Thomson (Lord Kelvin) (1824–1907) asserted that the atom "is a piece of matter with shape, motion and laws of action, intelligible subjects of scientific investigation."[2] It went without saying that the "laws of action" would have to provide an explanation for the characteristic spectral emissions, such as the Balmer series, associated with atoms.

With Thomson's discovery of the negatively charged electron, physicists had for the first time a tangible component of the atom to work with. And, as might be predicted, Prout's idea that atoms were built up from some common entity took on new significance. For Prout, the common entity was hydrogen; in the years immediately following 1897, the common entity became the electron. Lots of electrons. In 1900, for example, George F. Fitz-Gerald (1851–1901) suggested that the hydrogen atom "consisted of some 500 electrons," and three years later, J. J. Thomson asserted that hydrogen "contains about a thousand electrons."[3]

There were two major problems with these early, many-electron models of the atom. First, atoms are electrically neutral. What provides the positive charge required to neutralize the negative charge of the electron? As there was no evidence on which to base a definitive response, physicists at first largely finessed this question. The second problem was the inherent instability of the many-electron models. Since atoms are stable, any tenable model of the atom must account for its stability. An atom made up of 1,000 electrons, each repelling all others, works against stability.

The question, "What is the origin of positive charge in the atom?" is accompanied by another question: "What is the form of the positive charge and where is it located relative to the negative charge?" In 1902, Lord Kelvin proposed that the atom consisted of a sphere of positive charge in which the electrons were embedded. In the following year, J. J. Thomson elaborated on Kelvin's idea by considering the stability of such an arrangement and, perhaps because he was so eager to find a suitable model for the atom, he found what he considered a hint of stability. In any event, the name of Thomson has become associated with the "plum pudding" model: negative electrons leavening a spherical batter of positive charge.

In 1906, Thomson made perhaps his greatest contribution to the pursuit of an atomic model. With several lines of reasoning, Thomson concluded that the number of electrons in an atom was approximately equal to an atom's atomic weight. On this basis, there would be only one electron in a hydrogen atom. A principal line of reasoning employed by Thomson involved hydrogen itself; namely, he derived a theoretical expression for the index of refraction for a monoatomic gas and when his result was compared with experimental data for hydrogen, the result suggested that the number of electrons per atom of hydrogen must be approximately equal to one.

The plum pudding model, a batter of positive charge with minute negative currants embedded in it, appeared to be consistent with experiments which showed that a beam of electrons could pass undeflected through a thin metallic foil. In other words, one might conclude, as Philipp Lenard (1862–1947) did in 1903, that the atom was mostly empty space. These data as well as the larger question about the inner structure of the atom prompted a most provocative line of experimentation by Ernest Rutherford (1871–1937). Manchester University was the site of these historical experiments, which Rutherford initiated soon after he arrived in 1907 to assume his responsibilities as Langworthy Professor of Physics.

Rutherford liked alpha particles. After all, he had discovered them in 1898. In 1908 he established that the alpha particle carried a double positive charge. Long before he had the experimental proof, Rutherford seemed to know that the alpha particle was a doubly charged particle associated with the helium atom.

Rutherford and his assistant Hans Geiger directed a well-defined beam of alpha particles at thin foils of aluminum and gold.

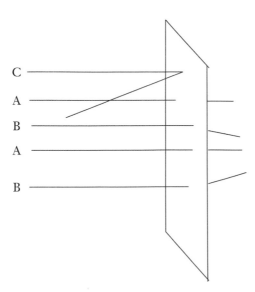

Figure 4.1 Alpha particles incident, from the left, on a gold foil. Most particles, like the A's, pass directly through the foil. A few particles, like the B's, are slightly deflected. A very few particles, like C, appear to bounce off the gold foil.

Most of the alpha particles passed straight through the foil (see Figure 4.1, label A), but some of them were scattered through a small angle (Figure 4.1, label B), especially from the foils composed of gold atoms. What in the atom of gold with its misty positive cloud and its tiny electrons could scatter the more massive, fast-moving alpha particles? Rutherford made a suggestion to Ernest Marsden, an undergraduate who was helping Geiger. Rutherford's suggestion went something like this: "Why don't you see if some alpha particles are scattered at large angles" (Figure 4.1, label C)? With Geiger looking on, the young Marsden pursued Rutherford's suggestion. The results were astounding: some alpha particles actually bounced off the gold foil in the general direction from whence they came. The effect was small: only about one alpha particle in 8,000 was reflected by the foil. But the implication of this small effect was clear: The alpha particle was hitting something substantial in the atom.

Rutherford published the results of these scattering experiments in mid-1909, and it seemed as if publication of the discovery of the nuclear atom would soon follow. But the plum pudding model remained the working model of the atom. Through the rest of 1909 and most of 1910, Rutherford pondered.

We cannot follow the details of Rutherford's ruminations exactly, but by late 1909, in an address to the British Association for the Advancement of Science, he accounted for the change of direction of alpha particles by iterating the unavoidable conclusion that "the atom is the seat of an intense electric field." At some point, Rutherford began to imagine a single encounter between something in the atom and the alpha particle. Whatever the intellectual path followed, we know that by the end of 1910, Rutherford's new conceptualization of the atom was taking form. In early 1911, a happy Rutherford encountered Geiger and announced, "I know what the atom looks like."[4]

Rutherford's atom consisted of a positively charged center some 10,000 times smaller than the atom itself. This center also carried most of the mass of the atom. For the gold atom, he found the charge at the center to be approximately 100 times the charge of the electron. Surrounding this center of positive charge were the electrons (see Figure 4.2). In March 1911, this new model of the atom was conveyed to the community of science. Later, in October 1912, Rutherford used the term nucleus for the first time.

Rutherford described his new model of the atom during a lecture he gave in Cambridge in the fall of 1911. J. J. Thomson listened to the lecture, but while the alpha-scattering data presented by Rutherford supported a "nuclear" model, Thomson did not. It may have been that another physicist, Niels Bohr, also heard this

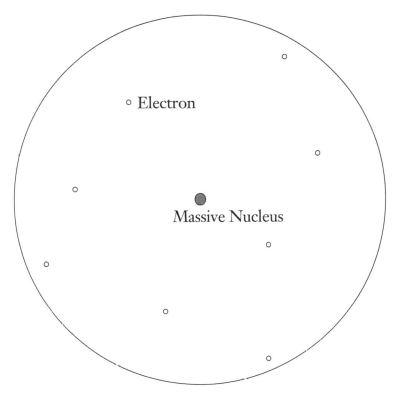

Figure 4.2 The nuclear atom of Rutherford. The electrons were distributed within a sphere surrounding the massive nucleus. No electron orbits were specified.

same lecture.[5] Bohr had recently begun his postdoctoral research at Cambridge University under the direction of Thomson.

Since Bohr was not particularly happy working in the crowded Cavendish Laboratory, he decided to go to Manchester and to Rutherford. With Rutherford's encouragement, Bohr made arrangements with Thomson to leave Cambridge and transfer to Manchester. Bohr's intention was to learn the experimental techniques of radioactivity. He arrived in Manchester in March of

1912 and he worked on the absorption of alpha particles by aluminum. However, his heart was not in experimentation. Other ideas were commanding his attention. By the time he left Manchester to return home four months later (July), these ideas were maturing.

More than any other person, more than any other physicist, Bohr was the guiding spirit of the quantum revolution. Bohr was not one of the physicists who, in the mid-1920s, created quantum mechanics: Bohr was not adept at creating the formal mathematical structures that were required. Unlike most physicists, Bohr's reputation did not emanate entirely from the papers he wrote. Yet, Bohr's contribution to twentieth-century physics is acknowledged by most physicists as second only to Albert Einstein's.

Bohr's importance can be attributed first and foremost to his model of the hydrogen atom. In a series of three papers, now called the Trilogy, Bohr laid the foundation for a quantum theory of atomic structure. The Trilogy was published in *Philosophical Magazine* in 1913 and these three papers established Bohr's reputation.[6]

Hydrogen: one negative electron and a positive nucleus. How does it go together? With hydrogen as his focus, Bohr confronted directly the problems inherent in the Thomson model and the nuclear model of Rutherford. The first problem was that both Thomson's plum pudding and Rutherford's nuclear models were neither mechanically nor electromagnetically stable. Electrons moving within Thomson's positive cloud or around Rutherford's nucleus continuously lose energy through the radiation of electromagnetic energy. This energy loss would be catastrophic: atoms would collapse and cease to exist.

Following Rutherford, Bohr depicted the hydrogen atom to be "a positively charged nucleus of very small dimensions and an electron describing closed orbits around it." Bohr used the term "stationary orbits" in which, he asserted, "there is no energy radiation," thereby violating the established tenets of nineteenth-century electromagnetism. Bohr recognized that the beautifully conceived and well-confirmed laws of seventeenth-, eighteenth-, and nineteenth-century physics could not apply within the atom. With bold recklessness, he selectively ignored the established laws of physics. "Whatever the alteration in the laws of motion of the electron may be," Bohr wrote at the beginning of the first paper of the Trilogy, "it seems necessary to introduce . . . the elementary quantum of action." With this sentence, Bohr placed the dynamics of atoms on a new foundation—a quantum foundation.

The Bohr hydrogen atom was founded on new physical ideas in which quantum conditions dictated restrictions on the older laws of physics. Bohr acknowledged a force—an attractive electrical force—acting between the negative electron and the positive nucleus. Bohr also acknowledged that an electron in orbit around a nucleus possesses energy and in terms of the Newtonian laws of physics, he could express this energy. This was Bohr's starting point. But immediately, Bohr placed a quantum condition on the energy; namely, he asserted that only certain energies are permitted. He identified these allowed energies as states of the hydrogen atom.

The quantum condition restricted the possible energies as well as the possible orbits of the electron. Only certain orbits were allowed. The larger the energy state of the hydrogen atom, the bigger the orbit. We can picture Bohr's hydrogen atom as a family of discrete orbits surrounding a central nucleus. The smallest orbit has the smallest energy; the next larger orbit, has the next larger energy, and so on to larger orbits and larger energies. In Figure 4.3, the principal elements of the Bohr model of the hydrogen atom are shown.

The most stable configuration of the hydrogen atom is the state

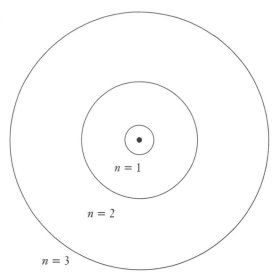

Figure 4.3 The Bohr model of the hydrogen atom. The radii of the electron orbits were determined by the model.

with the smallest energy. The hydrogen atom normally exists in the least-energy state; hence, the orbit associated with this state determines the size of the hydrogen atom. From his model, Bohr was able to calculate the radii of the various orbits and thus, from the radius of the smallest orbit, he could calculate the diameter of the hydrogen atom and hence, its size. The dimension of the hydrogen atom, as Bohr calculated it from his model, was 1.1 Å (1.1×10^{-8} cm). The experimentally determined diameter of the hydrogen atom is about 1 Å (1×10^{-8} cm)! Bohr had to be gratified and encouraged by this result: his model-generated diameter essentially agreed with the measured diameter of the hydrogen atom. Bohr was also able to calculate the ionization potential, which is related to the energy that is required to pull the electron entirely away from the nucleus. Once again there was essential agreement: thirteen volts (Bohr's calculated value) versus the experimental value cited by Bohr in the second paper of his Trilogy, eleven volts.

Another important result came out of Bohr's model. In Balmer's formula, which reproduced the wavelengths of the hydrogen spectrum, there appeared "the fundamental number of hydrogen" (b in Balmer's formula), which had the numerical value of 3,645.6 Å. As we shall see in Chapter 19, this number of Balmer's was generalized by Johannes Robert Rydberg in 1890 and became a very important physical constant. Bohr's treatment of the hydrogen atom transformed this constant from a simple number to a number that was expressed in terms of Planck's constant, the electron's mass, and its charge. In a March 6, 1913 letter to Rutherford, Bohr regarded this as an "enormous and unexpected development."[7] A little later (September 4, 1913), Arnold Sommerfeld wrote to Bohr: "calculating this constant is undoubtedly a great feat."[8]

Bohr brought one more basic idea to his treatment of the hydrogen atom, an amazing and a portentous idea: he proposed a mechanism to account for the spectrum of an atom. The idea was a rather straightforward application of his model. In keeping with the 1900 work of Max Planck, Bohr expressed the energy differences between two of his allowed states in the hydrogen atom in terms of Planck's constant and the frequency of light. For this process, Bohr assumed that the conservation of energy—a pillar of physical law from the nineteenth century—applied. In other words, the energy difference between a high energy state and a low energy state was exactly equal to the energy of the emitted light. This quantum transition was expressed as follows:

$$E_2 - E_1 = hf$$

where E_2 is the larger energy, h is Planck's constant, and f is the frequency of the emitted light.

Bohr wrote an explicit expression for the specific energies of the allowed states of the hydrogen atom. This expression is

$$E_n = -\frac{2\pi^2 me^4}{n^2 h^2}$$

where m = the mass of the electron = 8.85×10^{-28}; e = the electric charge of the electron = 4.7×10^{-10}; h = Planck's constant = 6.5×10^{-27}; and n = an integer = 1, 2, 3, . . . The negative sign in the energy expression simply indicates that the electron is energetically bound to the nucleus.

Bohr's expression for the energy difference between two energy states can be written in terms of the explicit expression just above. This takes the form:

$$E_{n_3} - E_{n_2} = \left[\frac{2\pi^2 me^4}{n_3^2 h^2}\right] - \left[\frac{2\pi^2 me^4}{n_2^2 h^2}\right]$$

$$= \frac{2\pi^2 me^4}{h^2}\left[\frac{1}{n_3^2} - \frac{1}{n_2^2}\right]$$

At some point during this process, Balmer's formula came to Bohr's attention, and it became "immediately clear." The clue for Bohr may well have been the appearance of n^2 in the denominators of both Balmer's formula,

$$\lambda = b\frac{m^2}{m^2 - n^2},$$

and his expression. In any event, it became clear to Bohr that the spectral lines in the Balmer series originated when the hydrogen atom emitted light whose energy, expressed as hf, was equal to the energy difference between two allowed energy states.

It is interesting to recognize that Bohr's creative imagination failed him in one significant instance. In 1905, a paper by Einstein was published in which he argued for a quantum view of light in which light was corpuscular in nature. Bohr, along with many other physicists, rejected Einstein's light-quantum idea and continued to do so until the early 1920s. So while Bohr adopted Planck's quantum idea, he rejected Einstein's. Had Bohr accepted a corpuscular view of light, he might have created a more physically picturesque model, thereby anticipating the form his model took within a few years of the 1913 publication. For example, later in 1913, an illustration of Bohr's model appeared that subsequently was used by Arnold Sommerfeld in his 1919 textbook on atomic structure and atomic spectra—the primary reference in the field for a period of a dozen or more years. Sommerfeld wrote that the diagram "summarizes" the Bohr model of the hydrogen atom. Sommerfeld's figure, similar to that in Figure 4.4, along with the energy-level diagram in Figure 4.5, are essentially the diagrams used by contemporary physicists to represent quantum transitions. Yet, Sommerfeld himself vacillated about the corpuscular view of light. Although he cited Einstein's view, he did not come out clearly in favor of it.

Bohr's unwillingness (along with his contemporaries) to accept Einstein's powerful idea notwithstanding, his accomplishment was truly magnificent. He created the first theoretical model of an atom that gave quantitative results for its physical properties: size and ionization potential. He created the first theoretical model of an atom that provided a mechanism for the origin of spectral lines, and it also gave quantitative results. In all cases, the numbers calculated from Bohr's model were in close agreement with experimentally measured numbers.

Despite the success of Bohr's model, many physicists took an intellectual step back from Bohr's ideas. Otto Stern threatened to leave physics if "that crazy model of Bohr" turned out to be correct.[9] (In Chapter 11 we'll see that Stern remained in physics.) As would be expected, older physicists were more likely to eschew

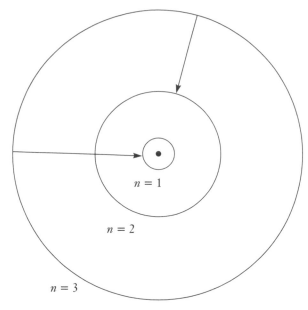

Figure 4.4 A diagram of the Bohr model of the hydrogen atom as shown by Arnold Sommerfeld in *Atomic Structure and Spectral Lines*. The radii of the pictured orbits $n = 2$ and $n = 3$ are four times and nine times larger than the radius $n = 1$. Orbits for $n = 4, 5, 6$, and so on are even larger. Spectral lines originate when the atom passes from one energy state, $n = 3$, to others, $n = 2$ and $n = 1$, as pictured.

Bohr's model and voice their confidence in an older, more familiar physics. Changing one's commitment from one conceptual scheme to another can involve an intellectual and emotional uprooting comparable to changing religions.

Even those physicists who recognized the significance of Bohr's achievement were aware of the fundamental challenge the Bohr model made to conventional ideas. For example, all science operates on the basis of cause and effect, that is, causality. Even before Bohr's paper appeared in *Philosophical Magazine*, Rutherford, who had seen a draft of the paper, raised a causality question: When an

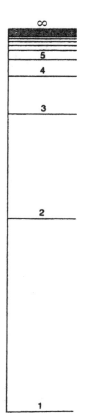

Figure 4.5 The energy-level diagram from the Bohr model provides an explanation of the Balmer series. Not drawn to scale.

electron leaves one energy state, how does it know where it is going? Rutherford saw this as a "grave difficulty."

The reaction to Bohr's model is understandable. Bohr's atomic model was based on older laws of physics with quantum assertions added. As such, it was clearly a jumbled affair. But the model provided a pictorial explanation of the origin of spectral lines and from the model the wavelengths of the Balmer series could be calculated. The model failed for the next simplest atom, helium. Had

Bohr focused his attention on any atom other than hydrogen, he would have failed.

There was one additional shortcoming with Bohr's model. In 1891, Albert Abraham Michelson discovered that the first Balmer line, called H_α, with a wavelength of 6,562 Å, was not one bright line, but two lines with almost equal wavelengths, very close together. Their discovery was made possible by the highly refined optical method Michelson applied to his observation of the Balmer spectral lines. This doublet structure of the Balmer line H_α could not be accounted for by Bohr's model.

In spite of the shortcomings of Bohr's model, the legacy of Bohr's atom is pervasive. We still talk of energy states of atoms. Our basic understanding of atomic and molecular spectra rests upon Bohr's idea of quantum transitions between energy states. And it was the success of Bohr's hydrogen model that affirmed the need to develop a new physics for atoms.

After Bohr's Trilogy, his reputation grew to enormous proportions not so much as a consequence of the papers he wrote, but as a consequence of the influence he exerted on others. In 1925–26, quantum mechanics, a new physics that brought understanding to the world of the atom, was created. Bohr was the epicenter of this activity and young physicists from laboratories around the world came to Copenhagen to be guided by Bohr. In that great period of physics, wrote Victor Weisskopf, "Bohr and his men . . . touched . . . the nerve of the universe."[10]

Relativity Meets the Quantum in the Hydrogen Atom

Arnold Sommerfeld, 1916

There's a reason physicists are so successful with what they do, and that is they study the hydrogen atom and the helium ion and then they stop.

—Richard Feynman

The quantum idea Bohr used in his model of the hydrogen atom was born in 1900. In that year Max Planck (1858–1947), the leading physicist at the University of Berlin, described the universal constant h, called Planck's constant. This constant establishes the scale of quantum phenomena. The extreme smallness of h explains why quantum effects are masked except at atom-size scales.

Hot objects glow; that is, they radiate heat at long wavelengths and visible light at shorter wavelengths. In 1900, no physicist could account for the way energy radiated from a hot body was distributed across various wavelengths. Planck provided the explanation and, in the process, he launched the quantum revolution. With the quantum idea, he developed a physical theory that accounted precisely for the experimentally observed distribution of energy across different wavelengths. Planck was aware of the importance of what he had done. One of Planck's sons reported that during a walk in their neighborhood of Grunewald, a suburb of Berlin, his father told him that he had made a major discovery comparable, perhaps, to the discoveries of Newton.[1] At the same time, however, he was not comfortable with the break his work represented from past physics and tried, unsuccessfully, to position his work in the older tradition he was at home with. By contrast, Bohr, Planck's junior by thirty-four years, saw the need for the quantum idea and incorporated it in his model of the hydrogen atom.

A second major break with the past, the special theory of relativity, was the brainchild of Albert Einstein, a twenty-six-year-old patent clerk, in Bern, Switzerland in 1905. No twentieth-century scientist has been regarded with such awe and reverence, by non-scientists and scientists alike, as Albert Einstein. After he generalized the relativity theory in 1915 and after one of its predictions, the bending of light by a massive object, was confirmed by Arthur Eddington's eclipse expedition, Einstein became a world celebrity.

The quantum idea and the implications of relativity theory first met through the work of Arnold Sommerfeld in 1915. Sommerfeld (1868–1951), a professor of physics at the University of Munich, was well known for his contributions to physical research. He was also a renowned teacher. Many famous physicists of the twentieth century, including Wolfgang Pauli, Werner Heisenberg, and Hans Bethe studied under Sommerfeld and scores of physics students used his numerous textbooks. His book *Atomic Structure and Spectral Lines* (1919) was a primary reference for physicists interested in the atom.

Although the Bohr model of the hydrogen atom was a tremendous triumph, Bohr's ideas were not immediately accepted; in fact, prominent physicists expressed considerable skepticism. Bohr's three papers, the Trilogy, were published in the spring and summer of 1913. In September of that year, Sommerfeld wrote a gracious letter to Bohr indicating his interest in the new model of the hydrogen atom. One difficulty with Bohr's model was its limitation to hydrogen. The model failed for the next simplest atom, helium. Furthermore, the ideas even failed to provide a complete explanation of hydrogen. At first blush, Bohr's theory appeared to account for the spectral lines of the Balmer series; in addition, the theory accurately predicted other spectral series of the hydrogen atom, seen in the infrared and ultraviolet. However, when the 6,562 Å Balmer line, H_α, was observed closely, it had fine structure; that is, it was not a single line, but a doublet—two lines close together. Bohr's 1913 theory could not account for this added complexity. This was the challenge that captivated Sommerfeld and he set out to see if he could provide an explanation for this complexity.

The nuclear model of the atom, as envisioned by Rutherford and Bohr, had much in common with the solar system. In each there is a massive core that exerts a controlling influence over less massive satellites orbiting around the central core. In both the solar system and the atom, the force between the central core and the orbiting satellites decreases as the square of their separation. In the case of the solar system, it was Johannes Kepler, early in the seventeenth century, who first allowed hard data—data he knew to be accurate—to sit in judgment on his speculations about the orbits of the Sun's planets.

When Kepler began the major work for which he became known, the structure of the solar system was a hot topic of intellectual and emotional debate. In accord with Copernicus, Kepler believed that the planets revolved around the Sun—an unpopular opinion in 1600. The prevailing and orthodox view was that the motionless Earth occupied the privileged central position in the universe with the Sun and planets orbiting around it.

Early in 1600, Kepler joined the astronomer Tycho Brahe at his observatory in Prague. Soon thereafter Kepler became consumed with the problem of establishing an orbit for the planet Mars. For this effort, Kepler had excellent data: accurate observations of Mars's position at various times over a period of years. These observations, made by Tycho Brahe, were the most accurate positional data for Mars that were available. The same orthodoxy that put the planets and the Sun in motion around the central Earth also held that the only appropriate orbit for celestial bodies was a circle. Kepler was a captive of this orthodoxy so, along with Copernicus, he put the Earth and the other planets in perfect circular orbits about the Sun. For most of nine years, Kepler tried unsuccessfully to fit Brahe's data for Mars into a circular orbit. Failure followed failure. Finally, in a fit of desperation, Kepler abandoned the circular constraint, and soon he was able to establish an orbit for the planet Mars consistent with Brahe's observations. The resulting orbit was an ellipse. This result later became generalized into Kepler's first law of planetary motion: Planets move in elliptical orbits with the Sun at one focus of the ellipse.

Like Kepler, Bohr assumed circular orbits for the electron's motion around the nucleus; unlike Kepler, however, Bohr was not guided by orthodoxy, but by reasons of simplicity. But what about those spectral lines which, upon close scrutiny, are not one line, but two . . . or more? This is where Sommerfeld enters the story. Sommerfeld generalized the Bohr model by considering the more general orbit—the ellipse. Actually, the ellipse is the more likely orbit for an electron moving under the influence of the force exerted on it by the nucleus. The same is true for the orbital motion of the planets. A planet *can* have a circular orbit, but the condition for circularity is much more specialized and hence more unlikely. So Sommerfeld relaxed the specific condition required for a circular orbit and considered an electron moving along an elliptical path.

The ellipse gave Sommerfeld additional freedom. The circle has one radius; the ellipse has two (see Figure 5.1). We say that the ellipse has a major axis, a, and a minor axis, b. In the Bohr model, the radius of the circular orbit is determined by the quantum number n. For an elliptical orbit, two quantum numbers are needed: one to specify the major axis and a second to specify the

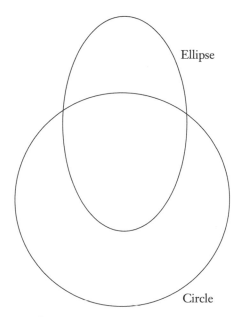

Figure 5.1 Sommerfeld relaxed the specific condition required for a circular orbit in the Bohr model and also considered an electron moving along an elliptical path.

minor axis. Bohr's quantum number specified the former and a new quantum number, *k*, took care of the latter. Sommerfeld set out to generalize Bohr's quantum condition for electron orbits in terms of the quantum numbers *n* and *k*.

Since the quantum domain is characterized by discontinuity with specific allowed states, quantum numbers, which effectively label the states, are the essence of the quantum domain. Multiples of a particular quantum number give the numerical values of physical properties to a quantum system like the atom. In the world we observe around us, physical properties like a person's weight take on a continuous range of values. For example, a human's body weight does not make a quantum jump from one weight-state, say 150 pounds, to another weight-state, say 160

pounds; individuals' weights are found at all weight values between 150 and 160 pounds. Weight varies continuously. But in
the domain of the atom, quantum principles govern and physical
properties like the energy states of an atom are quantized. Atomic
energy-state values vary discontinuously with leaps from E_1 to E_2,
from E_2 to E_3, and so on, and each state is labeled by a quantum
number.

In Bohr's model of the hydrogen atom, the circular orbits were
determined by the quantum number n; more accurately, by the
square of the quantum number n. No other orbits were allowed.
By changing the orbits from circles to ellipses, Sommerfeld introduced a second radius, which gave him another variable to play
with. So it was that Sommerfeld generalized Bohr's quantum condition for electron orbits in terms of the two quantum numbers: n
and k. His analysis led him to establish a relationship between the
two quantum numbers; namely, the quantum number n set the
upper limit on the quantum number k, but k could have smaller
values as follows:

when $n = 1$, the quantum number $k = 1$;
when $n = 2$, the quantum number $k = 2$ and 1;
when $n = 3$, the quantum number $k = 3, 2$, and 1;

and so on for higher values of n.

In terms of these quantum numbers, what did the new orbits
for the hydrogen atom look like?

For $n = 1, k = 1$: the orbit is a circle ($a = b$) with $r = r_1$;
for $n = 2, k = 2$: the orbit is a circle with a radius four times
larger than $n = 1$;
for $n = 2, k = 1$: the orbit is an ellipse with $a = 4r_1$ and $b = 2r_1$;
for $n = 3, k = 3$: the orbit is a circle with a radius nine times
larger than $n = 1$;

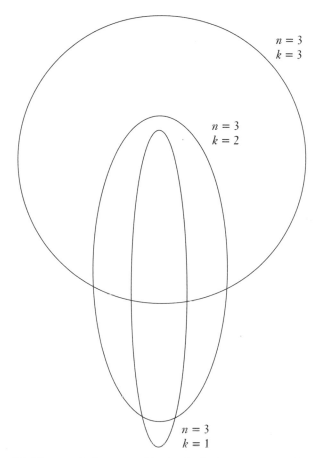

Figure 5.2 The $n = 3$ electron orbits in the hydrogen atom as determined by the quantum numbers n and k.

for $n = 3$, $k = 2$: the orbit is an ellipse with $a = 9r_1$ and $b = 6r_1$;

for $n = 3$, $k = 1$: the orbit is an ellipse with $a = 9r_1$ and $b = 3r_1$;

and so on for higher quantum number n (see Figure 5.2).

These were Sommerfeld's orbits for the hydrogen atom. When the atom is in the $n = 2$ state, the electron can move in either the circular orbit ($k = 2$) or the elliptical orbit ($k = 1$); in the $n = 3$ state, it can move in three different orbits.

And now, at last, the theory of relativity comes on the scene. According to relativity theory, the mass of an object increases with its speed. Granted, this increase does not become appreciable until the speed gets very large. But as the electron sweeps in toward the nucleus on its elliptical path, the speed increases significantly and with that speed increase there is an attending increase in the mass of the electron. The relativistic effect of changing mass cannot be ignored. The mass increase that occurs when the electron moves through its orbital perigee (the point closest to the nucleus) gives rise to a slight energy change associated with the elliptical orbit and two energy states result. In Bohr's model, the energy state labeled by $n = 2$ had one energy; in Sommerfeld's modified model, the energy state $n = 2$ has two orbits, $k = 2$ and $k = 1$, and due to the relativistic mass increase in the $k = 1$ elliptical orbit, its energy is slightly different from the circular $k = 2$ orbit. This slight energy difference gives rise to a set of two lines for each Balmer transition and thereby provides an explanation for the doublet appearance of the H_α line, and other lines as well (see Figure 5.3).

Sommerfeld's work was based on Bohr's model of the hydrogen atom. In this work, he brought relativity theory and the quantum idea together and was able to account for the fine details of the hydrogen spectrum. After Sommerfeld's paper on this work was published in 1916, he received a letter from Niels Bohr. In it Bohr wrote, "I do not believe ever to have read anything with more joy than your beautiful work."[2]

Like Bohr's model of the hydrogen atom, Sommerfeld's theory flowered only briefly. The creation of quantum mechanics and the discovery of electron spin, both in 1925, followed by Paul Dirac's theory in 1928, provided a solid theory-based underpinning for

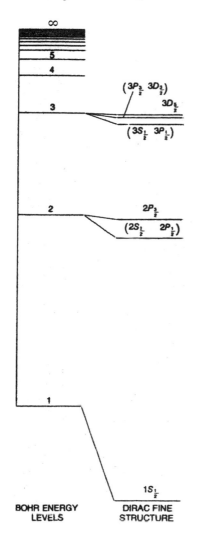

∞

5
4

(3P₃/₂ 3D₃/₂)

3D₅/₂

3

(3S₁/₂ 3P₁/₂)

2

2P₃/₂

(2S₁/₂ 2P₁/₂)

1

1S₁/₂

**BOHR ENERGY
LEVELS**

**DIRAC FINE
STRUCTURE**

Figure 5.3 The hydrogen energy states according to Bohr, then Sommerfeld and Dirac. Not drawn to scale.

the fine structure of the hydrogen spectrum. Nonetheless, Sommerfeld's successful extension of the Bohr model helped to strengthen the efficacy of Bohr's ideas, hence, after 1916, the quantum conditions used in atomic physics were called the Bohr-Sommerfeld rules.

6

The Fine-Structure Constant: A Strange Number with Universal Significance

Arnold Sommerfeld, 1916

With respect to [the fine structure constant] we are in the rather humiliating position of people who have to wrap a piece of string around a cylinder to determine pi.

—Edward M. Purcell

There are many constants in physics, but there are only a few fundamental constants and they are special because they have universal significance. Although the fine-structure constant is the subject of this chapter, let's begin with six other basic constants of nature. These six, with their current best values, are:

the gravitational constant, $G = 6.67259 \times 10^{-11}$ meter^3/kilogram second (m^3/kg s);
the speed of light, $c = 2.99792458 \times 10^8$ meter/second (m/s);
Planck's constant, $h = 6.6260755 \times 10^{-34}$ joule second (J s);
the electron's charge, $e = 1.60217733 \times 10^{-19}$ coulomb (C);
the electron's mass, $m_e = 9.1093897 \times 10^{-31}$ kilogram (kg);
the proton's mass, $m_p = 1.6726231 \times 10^{-27}$ kilogram (kg).

Immediately, one thing stands out: these constants are not approximate numbers. They are known with extreme precision. There is an uncertainty (not shown explicitly above) in each of the last digits, for example the last digit in Planck's constant or the last digit in the electron's charge. In each case, however, the digits up to the last are real. The precision of these constants is not an accident. There are reasons to know these numbers with the greatest precision possible; thus, ingenious experiments have been and are being designed to measure these constants with ever greater precision.

The importance of these constants is manifold. For example, are they really constants? Or have they changed over the life of the universe? Are they changing even now? If they are changing, the change over the course of human history would be minuscule and our only chance of detecting any change is by pushing our knowledge of these constants to ever greater precision.

Enormous significance derives from the universality of these constants. For every galaxy in the universe, in whatever way the stars are distributed throughout it, the unseen force of gravity extends across the light years, binds star with star, and gives the galaxy its shape. No current evidence challenges the conclusion that the gravitational force is the same everywhere. The gravitational constant, $G = 6.67259 \times 10^{11}$ m^3/kg s, determines the strength of the attractive force that binds the Earth to the Sun and holds the Milky Way together. For each star in the universe, in whatever galaxy it resides, the light emanating from it travels at the same speed, $c = 2.99792458 \times 10^8$ m/s. For each atom in the universe, in whatever material environment it exists, its size and its behavior is determined by Planck's constant, $h = 6.6260755 \times 10^{-34}$ Js. Every electron in the universe carries the same charge and has the same mass. The same holds true for the proton.

The actual numerical values of these constants depend on the units used to express them. If, for example, we used the foot rather than the meter for the unit of length, the gravitational constant, G, would have the value 2.35640×10^{-9} ft^3/kg s. So the value of these constants depends on the units employed. However, if we had a method of comparing units with an intelligent being from Galaxy X, we would find their numerical values for these constants agree with ours.

The first three constants listed above, G, c, and h, have their origins in four great theories: Newton's theory of gravitation, Maxwell's electromagnetic theory, Einstein's theory of relativity, and quantum mechanics. The actual values of these constants have meaning in terms of these theories. The second three constants have their origin in the nature of matter itself. These constants determine the nature of our material world. Suppose, for example, the charge on the electron were two times larger. The hydrogen atom would then be one-quarter its present size. All other atoms would be likewise scaled down in size and standard eight-foot ceilings would become two-foot ceilings.

Why the latter three constants, the charge and masses of the electron and proton, have their particular values is not understood. Why is the mass of the electron $9.1093897 \times 10^{-31}$ kg and not something else? This is an active area of investigation. One line of reasoning provides at least a rationale: if their observed values were much different, the human species would not be here to observe them. This reasoning purports to show that certain physical and biological processes are required for life, so these constants must have the particular values they have to support such forms of life. In other words, a standard two-foot ceiling could not have little Bohrs and Sommerfelds working under them.

The fine-structure constant is another fundamental constant, which first appeared in Sommerfeld's work on the hydrogen atom. Its value is $\alpha = 0.00729735308$. More often, this constant is written as follows:

$$\alpha = \frac{1}{137.0359895}.$$

The fine-structure constant is more than just a number. This constant can be expressed in terms of other constants; namely,

$$\alpha = \frac{e^2}{2\varepsilon_0 hc}.$$

The fine-structure constant derives its name from its origin. It first appeared in Sommerfeld's work to explain the fine details of the hydrogen spectrum. Recall that Bohr's model of the hydrogen atom provided a mechanism for the origin of spectral lines; namely, quantum jumps from one energy state to another. Bohr's model successfully accounted for the principal features of the hydrogen spectrum. On closer examination, however, the H_α spectral line in the hydrogen spectrum was, in fact, a set of distinct lines. Sommerfeld accounted for this fine structure by complementing Bohr's circular orbits with elliptical orbits. By treating the electron in these elliptical orbits relativistically, Sommerfeld accounted for the observed fine structure. Since Sommerfeld expressed the energy states of the hydrogen atom in terms of the constant α, it came to be called the fine-structure constant.

This constant explains far more than the appearance of the hydrogen atom's spectrum, however. The fine-structure constant is recognized as one of the most important constants in physics. We know, for example, that the fine-structure constant is a measure of the strength of the interaction between photons and electrons. Thus, this constant will appear in all situations that reveal quantum and relativistic properties of electrically charged particles. If electrons and light did not interact, the fine-structure constant would be zero.

The fine-structure constant is endowed with special significance because it is dimensionless.[1] In this regard, α is like the dimensionless constant π, the ratio of a circle's circumference to its diameter: $2\pi r/2r = 3.14159. \ldots$. The fine-structure constant, α, is unlike the speed of light, which has units of m/s, or the charge of the electron, which is measured in coulombs. The fine-structure constant, α, is independent of units: all intelligent beings, everywhere in the universe, share the same numerical value for the fine-structure constant. Citizens on a planet in another galaxy would express the speed of light as a different numerical value because their units would differ from ours, but the fine-structure constant would be identical—1/137.0359895.

Another characteristic adding to the mystique of the fine-structure constant is its ubiquitous nature. It emerges from a number of distinct physical situations, each of which permits a rather precise evaluation of the value of this dimensionless constant. For example:

(a) from measurements of h/m_n (h is Planck's constant and m_n is the mass of the neutron) comes the value $\alpha = 1/137.03601082$;

(b) from measurements of the alternating current Josephson effect at a superconducting junction comes the value $\alpha = 1/137.0359770$;

(c) from measurements of e^2/h (e is the charge of the electron) in the quantum Hall effect comes the value $\alpha = 1/137.0360037$;

(d) from measurements of the magnetic moment of the electron and positron comes the value $\alpha = 1/137.03599993$;

(e) from measurements of the energy states of the muonium "atom" consisting of an electron in orbit around a muon[2] where the muon serves as the nucleus comes the value $\alpha = 1/137.0359940$;

(f) from measurements of the helium spectrum comes the value $\alpha = 1/137.035853$.[3]

The details of the experiments that give rise to the results above are not important for our considerations. What is important is that each of the measurements (which contain experimental uncertainties that are not shown) arise from the different physical systems and, as is apparent, the values of the fine-structure constant emerging from these experiments are essentially in agreement.

The quest for ever greater precision and accuracy of experimental measurements is a staple not only of physics, but of any other science that has conceptual models or theoretical systems. The test of all science is experiment. Conceptual models can fascinate the minds of scientists and stir their emotions. They may regard a theoretical idea as so beautiful and so provocative that they cling to it with the eager anticipation that it will bring fresh insights to the secrets of nature. But, if the model or idea does not provide the opportunity for experiments to test its validity, even the most stubborn scientist will eventually abandon the idea regardless of its inherent charm. And here is where the fine-structure constant comes to the fore.

Quantum electrodynamics (QED) is one of the most successful, unifying theories of physics.In fact, the theory of QED underlies all the experiments I have just listed. Furthermore, with QED and the fine-structure constant, physicists can predict the values of many physical parameters to a high level of precision. For these reasons, QED is highly regarded by physicists. Nonetheless, QED, like all theories of physics, is always vulnerable. Since the theory of QED underlies all the various experiments shown above, the measured values of the fine-structure constant from these different experiments should be the same. If these experiments revealed different values of α, even slightly different values, questions as to the validity of QED would automatically follow. That's the way physics and other quantitative sciences work.

The fine-structure constant is badly named. Although it did originate from a study of hydrogen fine structure, the constant's significance far transcends its origin. We have no theory that allows us either to predict or to calculate the fine-structure constant. Edward M. Purcell wrote, "With respect to α we are in the rather humiliating position of people who have to wrap a string around a cylinder to determine π."[4] Attempts to find meaning in the fine-structure constant have stimulated flights of fancy and mystical musings. In turn, these abortive attempts were the target of what has been called "arguably the best physics joke ever to slip by an editor of a first-rate physics journal."[5] The joke was perpetrated by three young postdoctoral fellows at the Cavendish Laboratory. One of the comedians was Hans Bethe who, eleven years later, was to head the theoretical division at Los Alamos Laboratory. Bethe and his two cohorts wrote a paper purporting to show that the fine-structure constant was exactly 1/137. The paper was nonsense and the editor of *Naturwissenschaften* published it unawares. He was understandably furious and demanded an apology. On March 1, 1931, an apology appeared. "The Note," they wrote, "was not meant to be taken seriously. It was intended to characterize a certain class of papers in theoretical physics of recent years which are purely speculative and based on spurious numerical agreements."[6]

The meaning of the fine-structure constant will not come through "spurious numerical agreements." Perhaps if we receive an extraterrestrial message that we decipher as 137.0359895, we shall wonder whether the understanding of physicists somewhere else in the universe is deeper than our own. If such a provocative signal is received and if we haven't learned how to derive and interpret this constant from basic principles, we might convince the intelligent beings out there to tell us what's up.

7

The Birth of Quantum Mechanics: The Hydrogen Atom Answers the "Crucial Question"

Werner Heisenberg and Wolfgang Pauli, 1925–26 • Paul Dirac, 1925–26

> We now come to the crucial question for the whole new theory: Is it able to account for the properties of the hydrogen atom?
>
> —Max Born

In the 1960s, physicists were asked to identify the greatest physicist who ever lived. Competing for the top spot were Isaac Newton and Albert Einstein. Coming up third was a scientist who lived in the third century BC, Archimedes. For many reasons, Newton and Einstein qualify for this honor. Newton essentially brought heaven and Earth together by demonstrating convincingly that the same laws applied to each, and by this he established the universality of physical laws. Einstein crafted the special and general theories of relativity, fundamentally changing the spatial and temporal stage on which the events of the physical universe are played out. Relativity theory was one of two twentieth-century theories that revolutionized physics; the other was quantum mechanics.

If the same physicists who named Newton and Einstein as the all-time greatest physicists were asked what theory of physics has had the most pervasive influence on physics and physicists' view of the world, they would undoubtedly name quantum mechanics. For although relativity theory rebuilt the space-time stage on which nature's events unfold, quantum mechanics changed in a most fundamental way the character of physical laws obeyed by these events.

The quantum idea made its appearance in 1900 at the hands of Max Planck. Its implications were recognized by Einstein who, in recalling his attempts shortly after 1900 to understand the quantum theory, remarked, "It was as if the ground were pulled out from under one."[1] The solid footing underlying the work of physicists had indeed been ravaged by Planck's quantum idea and for the next twenty-five years a mood of crisis hung over the discipline, with anomalies following on the heels of paradoxes.

Evidence, in the form of hard data amassed in the laboratory, is the grist of physics. Between 1900 and 1925, the evidence that accumulated revealed behaviors at the atomic level that stumped physicists. On the one hand, behaviors were observed that were *localized* in character. For example, Einstein demonstrated in 1905 that light, long known for exhibiting wave behaviors, was found to interact with atoms and electrons as though it were a particle. In an even more dramatic way, Arthur Compton demonstrated in 1923 that light bounces off an electron as though the two were tiny billiard balls. Similarly, electrons were observed to strike a detector screen at a particular spot just as a pebble strikes the surface of a pond at a particular point. So both light and electrons exhibited a localized behavior characteristic of a particle.

On the other hand, both light and electrons behaved in ways that were *delocalized* in character. Light had long been regarded as a wave with parameters like wavelength and frequency providing an accurate account of diffraction and interference. In 1923, Louis de Broglie, inspired by Einstein's 1905 extension of particle-like properties to light, extended wave-like properties to the electron and other particles, thereby indicating the universality of wave-particle duality. Although de Broglie's daring idea could not be experimentally verified until 1927, it did provide a physical basis for one of the assumptions in Bohr's model of the hydrogen atom. Einstein said in a letter to H. A. Lorentz: "I believe [de Broglie's idea] is a first feeble ray of light on this worst of our physics enigmas."[2]

The dual wave-particle character of light and electrons was, as Einstein said, an enigma. Inherently, particles and waves are contradictory in nature. A pebble has a definite location; water ripples do not. A pebble follows a definite trajectory on its way to the water's surface and as it sinks below the surface; water ripples do not have a definite trajectory. Pebbles can be pushed to greater and greater speeds; waves cannot be pushed—waves have one speed that depends on the medium carrying the wave. The antithetical nature of localization, in the case for a particle, and delocalization, in the case for waves, defied unification. Once again, Einstein captured the conundrum facing physicists in 1924 when he wrote, "There are therefore now two theories of light, both indispensable—as one must admit today despite twenty years of tremendous effort on the part of theoretical physicists—without any logical connection."[3] Einstein could have written the same statement about particles.

Localization and delocalization express the ideas of discreteness and continuity. The quantum world with its atoms, electrons, and photons is characterized by discreteness. By contrast, the living world, that is, the world of our experience, is characterized by continuity. As we seek to understand the quantum world, we are handicapped because our imaginations are the products of our experience. To our way of thinking, localization (discreteness) and delocalization (continuity) are contradictory. Yet both the electron and light do what they do despite our conceptual chagrin. The electron is just the electron and light is just light. The challenge is to construct a conceptual bridge between these disparate worlds. In 1925 the young German physicist Werner Heisenberg began work on just such a bridge.

In the summer of 1925, Heisenberg wrote the first draft of his epoch-making paper. He was only twenty-three, but by that time, he had received his doctoral degree from the University of Munich, where he had worked under the direction of Arnold Sommerfeld; he had spent over a year in Göttingen working with Max Born; and he had spent several months in Copenhagen collaborating closely with Bohr. So, although Heisenberg was a young man, he had learned from and collaborated with some of the world's leading physicists.

Heisenberg was a brilliant student with superb theoretical insight and mathematical skills. He was not adept in the laboratory, however, and this shortcoming cast a pall over the seminar room in the Theoretical Physics Institute in Munich on July 23, 1923. Professors had gathered here to administer Heisenberg's oral exam, the final hurdle between him and his doctoral degree. In the room were Sommerfeld and Willy Wien, the head of experimental physics. The twenty-two-year-old Heisenberg breezed through the challenging theoretical questions. Then came Wien's turn to quiz the candidate. Heisenberg stumbled and was unable to answer Wien's questions. Sommerfeld and Wien had their differences and Wien did not support conferring the degree on Sommerfeld's prize student. After heated discussions concerning the relative merits of theoretical and experimental physics, a compromise was reached: Heisenberg was given the second lowest passing grade. Humiliated, Heisenberg soon left Munich and headed for Göttingen, where he was to assume the position of assistant to Max Born, director of the Institute of Theoretical Physics.

When Heisenberg arrived in Göttingen in October 1922, he found himself in a different intellectual environment from that of Munich. Göttingen was the home of many world-class mathematicians like Richard Courant and David Hilbert. The study of physics in Göttingen was enveloped in a mathematical formalism. Sommerfeld also took a mathematical approach, but grounded it in a thorough knowledge of physical details, and thus Sommerfeld's mathematics was pragmatic. By contrast, Born was interested in mathematical methods and did not possess Sommerfeld's knowledge of the physical phenomena; thus, there was an axiomatic character to Born's mathematical physics and a fascination with mathematical subtleties. Born wanted mathematical, not physical proof. As Heisenberg said later, "In some ways, mathematics formed the whole spirit of Göttingen."[4]

Heisenberg took advantage of the Göttingen environment and learned new mathematics by attending classes given by Hilbert and Courant. As for physics, Heisenberg struggled long and hard to extend the quantum principles to the second simplest atom, helium. Helium, however, is not hydrogen and the number-two atom did not succumb to Heisenberg's efforts; furthermore, the principal ideas he developed in his effort to tame helium were harshly criticized by both Wolfgang Pauli and Bohr.

Heisenberg had attracted the attention of Bohr when Sommerfeld took his prize student to hear a lecture series by Bohr in Göttingen, June 12–22, 1922. As an outgrowth of an objection Heisenberg raised to one of Bohr's points, the two men took a walk and Bohr suggested that Heisenberg come to Copenhagen for a visit. Thus, in the spring of 1924, Heisenberg arrived at Bohr's institute in Copenhagen for a relatively brief visit. Things went so well that Bohr invited Heisenberg to return in the fall of 1924 to spend the winter semester. Born was reluctant to part with Heisenberg, but in April 1924, wrote to Bohr: "I shall, of course, miss him (he is charming, worthy, very bright man who has become very dear to my heart), but his interests precede mine, and your wish is decisive for me."[5]

Göttingen was different from Munich; Copenhagen was different from both. The influence of Bohr on the young Heisenberg cannot be exaggerated. It was Pauli who recognized that Heisenberg lacked what Bohr could supply. As Pauli expressed in a letter to Bohr, Heisenberg needed a more philosophical style of thinking about the physics of the atom in which ". . . a clear formulation of the basic assumptions and their connection with erstwhile theories" plays a central role.[6] Bohr was obsessed with clarity of thought, with understanding the physics of a problem before any attempt was made to develop a theoretical explanation. Later Heisenberg talked about his time with Bohr:

> Bohr never looked on problems from the mathematical point of view; he looked at them from the physical point of view. I learned more from Bohr than from anybody else that the new type of theoretical physics was almost more experimental than theoretical. That is, one had to cover the experimental situation by means of concepts which fitted. Later on one had to put the concepts into mathematical forms, but that was a more or less trivial process which had to be solved. But the primary thing here was that one had to find the words and concepts to describe a strange situation in physics that was very difficult to understand.[7]

When Heisenberg left Copenhagen in April 1925 to return to Göttingen, his days as a student were over. On April 21, Heisenberg wrote a letter to Bohr thanking him for his generous hospitality during his Copenhagen stay. "With respect to science," Heisenberg wrote, "the past half year was for me the most beautiful of my entire life as a student. I am almost sad about the fact that I must carry on wretchedly alone by myself in the future."[8] Heisenberg was about to embark on his own work. In this solitude, he would quickly create the first version of quantum mechanics.

With youth, commitment is to the future, not to past ideas.

This is why young physicists are able to make a clean break with the past. At the same time, evidence shows that a struggle often ensues in the mind of a young scientist who sees a new foundational principle that is at odds with those learned as a student and held dear by mentors and colleagues.

About one month before Heisenberg laid the formal foundation of quantum mechanics, he wrote a letter to Bohr in which he revealed, "Recently I have been occupied with the intensities [of spectral lines], notably in the case of hydrogen. . . . The present conditions are still not entirely sufficient to obtain the intensities."[9] The intellectual context underlying Heisenberg's remark was the Bohr-Sommerfeld model of the hydrogen atom; specifically, whether the light emitted in a spectral transition is bright or dim.

Why did the intensities of the hydrogen Balmer series occupy Heisenberg's mind? Part of the answer is the mood that was prevalent at that time: a deep sense of frustration gripped the minds of those few physicists who, in the fall of 1924 and into the spring of 1925, were vainly trying to put the physics of atoms on a rational quantum foundation. The fundamental contradiction that plagued atomic theory was clearly captured by remarks of Hendrik Kramers in his address to the Congress of German Scientists and Physicians on September 24, 1924:

> The theory of the hydrogen spectrum owes its success to the fact that the motions in the stationary states can be described in this case with the help of classical electrodynamics, or—to put it slightly differently—with the help of mechanics based on Coulomb's law of attraction. That this is possible, is not at all self-evident, since the fundamental postulates of Bohr are in open contradiction to classical electrodynamics. Thus the strange transitions between stationary states, whether stimulated by radiation or collisions, cannot be described in princi-

ple by classical methods, not even in the case of the hydrogen atom.[10]

With great intellectual exertion physicists crafted new ideas that, for a short time, appeared promising, but when analyzed further, fell short. Again and again, hopes soared, then plummeted.

In this gloomy context, the basis for a new approach began to take form in Heisenberg's thinking during the spring of 1925. The intensities of spectral transitions were a part of his new approach. He began to think about what was observable and what was inherently unobservable. Bohr's model of hydrogen was based on quantities that were unobservable. For example, the position of the electron in the hydrogen atom cannot be observed; therefore, Heisenberg set it aside as a fruitless idea. Likewise, electron orbits cannot be observed and so he ignored the idea of atomic orbits. In directing his attention to observable quantities, Heisenberg focused on the principal data that spawned the quantum revolution; namely, on the frequencies and intensities of spectral transitions. Spectral frequencies and intensities were experimentally observable; the coordinates of the electron and the orbits of the electron were experimentally unobservable. Heisenberg came to the conclusion that the physics of the atom should be expressed only in terms of observables.

On June 7, 1925, Heisenberg left Göttingen for what may be called a sick leave. His head was reeling not only because of the ideas that had taken form in his thinking, but also because of a pollen-induced allergy. During the next two weeks, on the rocky North Sea island of Helgoland, Heisenberg had the breakthrough that quickly led to the first version of quantum mechanics. In this creation, Heisenberg replaced position variables, measureable for planets, with the frequencies and intensities of spectral transitions, which are measureable for atoms. In other words, he discarded the position, x, of an electron in the atom and replaced it with a new descriptive variable relating to either the frequency or the intensity of spectral transitions between two energy states. After he developed some relationships for the new variables, he drew on basic Newtonian physics to describe how these variables change with time. Since an atom has the potential for many, many spectral transitions, each with different intensities, Heisenberg's quantum mechanics involved many, many equations.

By itself, Heisenberg's Helgoland creation was provocative, but incomplete. Sometime around June 20, Heisenberg left Helgoland and started back to Göttingen via Hamburg, where he visited Pauli. Heisenberg informed Pauli of his Helgoland results and was surprised to receive Pauli's encouragement rather than criticism. Over the next two weeks, Heisenberg worked hard to bring his Helgoland work to a form suitable for publication. On July 9, 1925, Heisenberg completed his manuscript.

At this stage, Heisenberg's manuscript could be judged neither right nor wrong. Heisenberg had tried to apply his new theory to the hydrogen atom, but was unable to do so. This failure left him discouraged and uncertain about his new approach. Nonetheless, his ideas, as laid out in the paper, were provocative. With no way to gauge the significance of his new ideas himself, Heisenberg realized that he needed an independent critique. There was no one better suited to the task than his friend, Wolfgang Pauli (see Figures 7.1 and 7.2).

From the beginning there was something special between Pauli and Heisenberg. They weren't friends in the social sense—their divergent lifestyles probably discouraged camaraderie after hours —but they were close professional friends bound together by a strong awareness of each other's intellectual prowess. Earlier, Pauli had come to Munich to pursue his doctoral work under the direction of Sommerfeld. While always respectful, Pauli early gained a reputation as a relentless critic of the intellectual products of his colleagues and associates. When Pauli graded the

Figure 7.1 Werner Heisenberg.

Figure 7.2 Wolfgang Pauli.

homework of the younger Heisenberg, he would tell Heisenberg, "You are a complete fool." Heisenberg would only work harder in response to Pauli's barbs.

So, given his uncertainty about the efficacy of his new theory, Heisenberg needed Pauli's judgment. Heisenberg transmitted his manuscript to Pauli in July 1925. In the covering letter, Heisenberg wrote that he wanted either to complete his manuscript and submit it for publication or to burn it. To help in this decision, he asked for sharp criticism. In a few days, Heisenberg had his response: Pauli was jubilant. At long last, he saw in Heisenberg's manuscript a rational basis for proceeding with the physics of the atom.

With Pauli's support, Heisenberg took his manuscript to Born in mid-July and asked him to decide whether the manuscript should be published. Born became so fascinated by Heisenberg's manuscript that he could "hardly sleep at night." His sleep was disturbed by a vague memory that refused to come into the light. Suddenly, after days of thinking devoted to Heisenberg's manuscript, the vague memory slipped out of the shadows. Born looked at Heisenberg's equations with their many, many spectral frequencies along with their many, many intensities and he recognized that they could be expressed in terms of matrix mathematics. Born sent Heisenberg's paper to the leading journal of physics, *Zeitschrift für Physik*, for publication and Born, along with Pasqual Jordan, began to develop the mathematics that would bring power to Heisenberg's ideas. In less than four months, two papers—the first by Born and Jordan and the second by Born, Heisenberg, and Jordan—were written in which the mathematical foundation of quantum mechanics was completed.

Heisenberg was scheduled to give a talk at the University of Cambridge on July 28, and after he gave his manuscript to Born for critique, he left for Cambridge. Heisenberg did not talk about his just completed work, but Ralph Fowler, who was in the audience, had heard the rumor of Heisenberg's recent work. When Fowler received galley proofs of Heisenberg's paper a few weeks later, he immediately sent them to his student Paul Dirac. Dirac was yet unknown in the community of physicists because he had started in electrical engineering and was only twenty-three. But Dirac was soon to become one of the elite group of physicists who laid the foundation of quantum mechanics.

Within a week, Dirac recognized that Heisenberg's paper represented a profound basis for bringing new understanding to the atom. Dirac's first effort with Heisenberg's paper was to introduce relativity into Heisenberg's treatment. This was unsuccessful. The breakthrough came during a Sunday walk in October when Dirac remembered some mathematics he had encountered in a course he had taken. This memory proved fruitful and Dirac was soon writing his paper, "The Fundamental Equations of Quantum Mechanics," which was published within three weeks of its completion in the November 7th issue of *Proceedings of the Royal Society* and soon became a classic. Dirac sent a handwritten copy of his paper to Heisenberg, who responded with congratulations on his "extraordinary beautiful paper on quantum mechanics."[11]

By the end of 1925, there were essentially two versions of quantum mechanics: Heisenberg's and Dirac's. The physical ideas were abstract and strange. The mathematics was foreign. Even the great Italian physicist Enrico Fermi found the ideas alien. What was needed was an application that would demonstrate the efficacy of the new quantum mechanics. The American physicist J. H. Van Vleck said, "I eagerly awaited to see if someone would show that the hydrogen atom would come out with the same energy levels as in Bohr's original theory, for otherwise the new theory would be a delusion."[12]

The hydrogen atom beckoned once again. Both Pauli and Dirac independently began the crucial test of the new theory:

would it explain the spectrum of the hydrogen atom? Pauli had great facility with mathematics, but he was disturbed by the gnawing worry that Born's complicated matrix mathematics would retard or even thwart the full development of the physical ideas, which, after all, were the most important. Pauli, in fact, had refused to help Born in his mathematical development. Born and Jordan, who embedded Heisenberg's ideas in the framework of matrix mathematics, attempted to apply the new formalism to the hydrogen atom, but they failed.

Despite his skepticism, Pauli learned the intricacies of matrix mathematics and applied the Heisenberg version of the new quantum mechanics to the hydrogen atom. In less than three weeks, Pauli obtained the same formula that Bohr had obtained in 1913, only this time the route to the formula was a coherent theory—the new theory of quantum mechanics. "Herewith," wrote Pauli, "it has been demonstrated that the Balmer terms come out directly from the new quantum mechanics."[13] So momentous was this demonstration that the skeptic Pauli became a believer in the matrix mathematical formulation of Born, Heisenberg, and Jordan.

Heisenberg was the first to hear the news. He was filled with joy and a bit of awe at Pauli's ability to accomplish the feat so quickly. Bohr heard the news almost as quickly as Heisenberg. "To my great joy," wrote Bohr in a November 13 letter to Pauli, "I heard . . . that you have succeeded in deriving the Balmer formula."[14] Bohr had reason to be happy; what he had done for the Balmer formula, Pauli had done for the new quantum mechanics.

Starting somewhat later, Dirac applied his version of the new quantum mechanics to the hydrogen atom. Dirac's approach to the hydrogen atom was very different and more general than Pauli's method and it too was successful in explaining the Balmer spectrum of the hydrogen atom. Dirac's paper on the hydrogen atom appeared in the *Proceedings* shortly before Pauli's paper appeared in *Zeitschrift für Physik* because of a faster publication track. Heisenberg was thrilled and wrote to Dirac in early 1926: "I congratulate you. I was quite excited as I read your work."[15]

During the fall semester of 1925, Born was in the United States lecturing at the Massachusetts Institute of Technology. He reminded his audience that "the explanation of the hydrogen spectrum was the first great success of Bohr's theory." However, he noted, "If the new theory failed here [to explain the hydrogen spectrum,] it would have to be abandoned in spite of its many conceptual advantages, but, as Pauli has shown, it stands the test successfully."[16]

With the hydrogen spectrum explained, the new theory of quantum mechanics had passed the crucial test. What remained were simply many, many details.

8

The Hydrogen Atom: Midwife to the Birth of Wave Mechanics

Erwin Schrödinger, 1926

In its special role as the simplest of all atoms, hydrogen has starred in some great episodes in the history of science.

—Daniel Kleppner

Numbers are so important in physics that real numbers alone are not enough: imaginary numbers are also used and often appear in the equations of physics. Physicists are enamored with numbers, so imaginary numbers are not that surprising; however, it is a bit surprising that an imaginary cat appears in the lore of physics. The cat belongs to Erwin Schrödinger. This imaginary cat was invented in 1935 as part of a thought experiment designed to probe the implications of the subject that Schrödinger himself helped to create—quantum mechanics. And because the interpretation of quantum mechanics has proven so provocative and continues to challenge our understanding, Schrödinger's imaginary cat still lives. Recently, the cat moved from the realm of the imagination into the laboratory. After we consider Schrödinger's role in the creation of quantum mechanics, we shall return to the curious cat.

From mid-1925 through the early months of 1926, events relating to quantum mechanics occurred in rapid succession. After years of intense efforts to develop a coherent quantum theory to explain the results of experimental measurements on atomic sys-

Figure 8.1 Erwin Schrödinger.

tems, the intellectual logjam broke open and there was an outpouring of epoch-making results. At the beginning of 1926, there were essentially two versions of quantum mechanics. On January 27, 1926, a manuscript containing a third version was received by the editors of the prestigious German physics journal *Annalen der Physik*. The author was Erwin Schrödinger, a professor of theoretical physics at the University of Zurich (see Figure 8.1). This paper was the first of six papers that Schrödinger, in an outburst of creative activity, wrote during the first six months of 1926. All the papers were fundamentally important, but the first paper was special. In this first paper, Schrödinger developed his version of quantum mechanics with the aid of the hydrogen atom.

Two stumbling blocks make it difficult to track the thought processes that led to his classic papers. First, unlike Heisenberg, who chronicled his thoughts in letters to Pauli and others, Schrödinger did not reveal himself as extensively in letters. He did keep journals, but key journals in this sequence have been lost. Schrödinger himself never provided a chronological account of how he arrived at his famous result. Second, Schrödinger's insights apparently came to him over such a compressed time period that attempts to reconstruct the evolution of his thought processes are difficult. We do, however, know a few things.

We know, for example, that in late 1925 Schrödinger read Louis de Broglie's paper in which the young de Broglie proposed that particles have an associated wavelength. Schrödinger sent a letter to Einstein on November 3 in which he wrote, "A few days ago I read with the greatest interest the ingenious thesis of Louis de Broglie, which I finally got hold of."[1] We also know that Peter Debye, Schrödinger's colleague in Zurich, suggested that Schrödinger give a talk on the de Broglie paper at the joint Zurich Federal Institute of Technology (E.T.H.)–University of Zürich colloquium. The colloquium probably occurred in November, shortly after Schrödinger posted his letter to Einstein. And we know further that Schrödinger went to work attempting to generalize de Broglie's concept of matter waves and treated it in the context of relativity theory. In fact, evidence suggests that Schrödinger first arrived at a relativistic form of his famous wave equation, which he rejected because it seemed at odds with available data. Had he known about electron spin, discovered a short time earlier, his relativistic result would have been very encouraging indeed. Unfortunately, no manuscript of Schrödinger's relativistic version exists. Still further, we know that Schrödinger was philosophically attracted to the continuity of waves and the way waves are represented mathematically. Finally, we know that Schrödinger left for a holiday shortly before Christmas 1925 and did not return until early January 1926.

The holiday at the Villa Herwig in Arosa was special. Accompanying Schrödinger was a young woman, whose identity remains a mystery. It was during this romantic interlude that Schrödinger arrived at his version of quantum mechanics, which, for a period of time, was called wave mechanics to distinguish it from Heisenberg's quantum mechanics. In January, Schrödinger submitted his manuscript for publication.

The first sentence of Schrödinger's classic paper reads as follows: "In this paper I wish to consider, first, the simplest case of the hydrogen atom, and show that the customary quantum conditions can be replaced by another postulate, in which the notion of 'whole numbers,' merely as such, is not introduced."[2] Two things about this sentence are noteworthy. First, an explanation of the hydrogen atom is clearly the objective of Schrödinger's wave mechanics. Second, in the development that follows this introductory sentence, quantum numbers ("whole numbers"), which appeared in Bohr's model of the hydrogen atom in a somewhat *ad hoc* fashion, appear as a natural consequence of Schrödinger's physical and mathematical approach.

The mathematics Schrödinger employed in his paper was familiar to physicists. He derived a wave equation, which has come to be called Schrödinger's equation. This famous equation is ubiquitous in modern science:

$$\frac{\partial^2 \psi}{\partial x^2} + \frac{\partial^2 \psi}{\partial y^2} + \frac{\partial^2 \psi}{\partial z^2} + \frac{8\pi^2 m}{h^2} \left[E - V(x, y, z) \right] \psi = 0.$$

Schrödinger set up this wave equation in a form appropriate for the hydrogen atom—a negatively charged electron orbiting around a positively charged proton (the nucleus). Of particular interest to Schrödinger was the variable E, or energy, in the above equation, which gave the energy states of the hydrogen atom. When Schrödinger solved the equation, out tumbled the energy states of the hydrogen atom: "Therefore the well-known Bohr energy-levels, corresponding to the Balmer terms are obtained."[3] As envisioned by Bohr, quantum transitions between these energy states, or Balmer terms as Schrödinger called them, have frequencies in agreement with the observed Balmer series. In the process of obtaining the energy states by his method, Schrödinger pointed out, "The essential thing seems to me to be, that the postulation of 'whole numbers' no longer enters into the quantum rules mysteriously."[4]

The Schrödinger paper was published in the March 13 issue of *Annalen der Physik*. With his usual penetrating perspicuity, Pauli wrote to Pasqual Jordan soon after Schrödinger's article appeared and said, "I feel that this paper is to be counted among the most important recent publications. Please read it carefully and with devotion."[5] Einstein wrote to Schrödinger on April 16, "the idea of your work springs from true genius," and ten days later he wrote, "I am convinced that you have made a decisive advance with your formulation of the quantum condition."[6]

Pauli's and Einstein's immediate positive reaction speaks of their remarkable intuition about the efficacy of new mathematical formalisms embracing new ideas. In fact, however, the full meaning and significance of Schrödinger's paper was not immediately obvious. When a wave equation is written, it is legitimate to ask, "What's waving?" In a water wave it is a succession of crests and troughs that "waves" the surface of the water. In the Schrödinger wave equation, what are the crests and troughs? This is equivalent to asking, what is the ψ that appears in Schrödinger's equation? Schrödinger himself made suggestions, but it was Max Born in June of 1926 who brought meaning to the ψ parameter and in so doing altered quantum physics in fundamental ways. Born said in his Nobel address in 1954, "The published work for which the honor of the Nobel prize for the year 1954 has been accorded to me does not contain the discovery of a new phenomenon of nature but, rather, the foundation of a new way of thinking about the phenomena of nature."[7]

In arriving at his interpretation of the ψ, called the wave function, Born was influenced by the way Einstein had tried to make the wave-particle duality of light comprehensible. In addition, Born was influenced by the atomic scattering experiments of his Göttingen colleague, James Franck, which for Born had a definite particle nature. "Every experiment by Franck and his assistants on electron collisions," said Born, "appeared to me as a new proof of the corpuscular nature of the electron."[8] When an electron is scattered from an atom, it can go in any direction depending on the nature of the collision. Born equated ψ, or rather $|\psi|^2$, with the probability that the electron would be scattered through a particular angle. At an angle where the quantity $|\psi|^2$ was large, the likelihood of detecting the electron at this angle was large. At an angle where $|\psi|^2$ was zero, the probability of detecting the electron there was zero.

When applied to the hydrogen atom, the magnitude of the quantity $|\psi|^2$ indicates where the electron is most likely to be found: where the value of $|\psi|^2$ is large, that's where the probability is largest for locating the electron. Figure 8.2 shows a plot of $|\psi|^2$ from the center of the hydrogen atom, r = 0, to larger distances from the atom's center. We see that $|\psi|^2$ has its largest value at a particular radial distance from the atom's center and this is where the electron spends most of its time. Schrödinger's quantum mechanics reveals this distance to be 0.529Å. The value 0.5Å agrees not only with Bohr's model of the hydrogen atom, it also agrees with the known size of the hydrogen atom, whose diameter is about 1Å. Schrodinger's theory and Born's interpretation are in agreement.

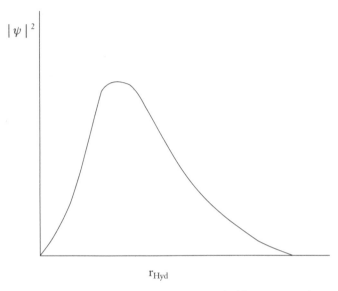

Figure 8.2 The magnitude of the quantity $|\psi|^2$ represents the probability of where the electron is likely to be found. It is a maximum at r_{Hyd}, which determines the radius of the hydrogen atom.

From this beginning, the quantity ψ came to be associated with probability, although of a different nature than physicists were accustomed to. Probability prior to Born was the result of ignorance. When a gambler throws dice, the details of the throw, details of the dice themselves, and details of the surface onto which the dice are thrown are not known precisely and this ignorance precludes predicting the outcome exactly. If all these details were known with absolute precision, the outcome of a dice throw could, in principle, be calculated. In our ignorance of these details, only the probability of a particular outcome, say a double six, can be given.

Born's probabilistic interpretation of ψ was fundamentally different. The probability of atomic events, say the decay of a single neutron, is *not* a probability of ignorance. Rather, the quantum laws that govern the decay process are inherently probabilistic. There are no details lacking; there is no ignorance. The quantum processes are probabilistic in their basic character. There is no way we can say when a neutron will decay; all we can say of a neutron is that in the next ten or twenty minutes there is a certain probability that the neutron will decay into a proton and an electron.

This interpretation of ψ brought wrenching change to physics. The laws of physics, prior to quantum mechanics, were laws of certainty. For example, physicists can predict absolutely when and where solar eclipses will occur. This certainty, however, is a consequence of scale: the Sun, Earth, and Moon are so massive that the inherent quantum behavior of these bodies is too small to affect their gross behaviors. However, at nature's most basic level, the domain of the basic building blocks of the material world, the laws of nature are laws of probability. Einstein could not accept this and his oft-quoted remark that God does not play dice reflects his rejection of the probabilistic interpretation of quantum behavior. Schrödinger himself never accepted the offspring of his own work, quantum probability. In August 1926, Schrödinger wrote to Willy Wien: "today I no longer like to assume with Born that an individual process . . . is 'absolutely random,' i.e., completely undetermined. I no longer believe today that this conception accomplishes much."[9] With no intention of doing so, Schrödinger had changed the nature of physical reality.

Whether or not physicists immediately accepted the implications of Schrödinger's quantum mechanics, they were pragmatic and recognized the opportunities inherent in the new mathematical formalism. "The Schrödinger theory came as a great relief," said George Uhlenbeck, the co-discoverer of electron spin.[10] Compared to the strange matrix mathematics, which conveyed Heisenberg's quantum mechanics, the mathematics of Schrödinger was familiar and physicists knew how to use it. And use it they did. Experimental results had been accumulating and Schrödinger's equation was applied to many atomic problems with stunning results.

Heisenberg and Schrödinger each created their theories for the exact same purpose: namely, to provide a theoretical framework for explaining the results of measurements made on atoms. For each author, success with the hydrogen atom proved the validity of the theory. Yet, though they were identically motivated, their results were radically different conceptually and mathematically. Heisenberg's quantum mechanics was expressed in terms of matrices, noncommuting quantities, and strange computation rules. Schrödinger's quantum mechanics started with a partial differential equation, the Schrödinger equation, which was solved by well-established methods. The Heisenberg quantum mechanics was conceptualized in terms of discreteness and the particle was the underlying image. The Schrödinger quantum mechanics was conceptualized in terms of continuity and the wave was the underlying image.

The differences between the two theories intrigued Schrödinger and in late February he set out to examine them. In March he sent a paper to *Annalen der Physik* that contained the results of his analysis. The first sentence of the paper sets the stage: "Considering the extraordinary differences between the two starting-points and the concepts of Heisenberg's quantum mechanics and of the theory that has been designated 'undulatory' or 'physical' mechanics, and has lately been described here, it is very strange that these two theories agree *with one another* with regard to the known facts."[11] In what followed, Schrödinger demonstrated the formal equivalence between the two theories—Heisenberg's and his own.

Equivalence notwithstanding, the differences between the Heisenberg and Schrödinger approaches prompted criticism from each about the other's ideas. On the one hand, Schrödinger rebelled against the abstractness of the Heisenberg approach and thought its lack of an intuitive basis or its lack of visualizability served to stifle creative applications. Schrödinger wrote of being "repelled" by the difficult methods of Heisenberg's approach. On the other hand, Heisenberg, who saw the atomic domain as one of discontinuity, lamented the continuity inherent in Schrödinger's approach. In June 1926, Heisenberg wrote to Pauli, "The more I think of the physical part of the Schrödinger theory, the more abominable I find it. What Schrödinger writes on the visualizability of his theory . . . I find rubbish."[12]

Heisenberg recognized the need to resolve the deep-seated differences between the two approaches. A face-to-face discussion with Schrödinger at a neutral site might accomplish a resolution. Where would be better for such a discussion than Copenhagen, where Bohr could participate? On September 11, 1926, Bohr wrote a letter to Schrödinger inviting him to Copenhagen to give a lecture on wave mechanics and to participate in "some discussions for the narrower circle of those who work here at the Institute, in which we can deal more deeply with the open questions of atomic theory."[13] The narrower circle would include, among others, Bohr, Heisenberg, and Dirac.

Schrödinger arrived in Copenhagen on October 1 and, as told by Heisenberg, "Bohr's discussions with Schrödinger began at the railway station and were continued daily from early morning until late at night. Schrödinger stayed at Bohr's house so that nothing could interrupt the conversations."[14] Schrödinger attacked quantum jumps; Bohr acknowledged that they could not be visualized, but defended them. Schrödinger wanted to know what went on in the atom; Bohr countered with the claim that concepts drawn from everyday experience could not be applied to the atom. On and on the discussions went until Schrödinger fell ill. Mrs. Bohr brought food and tea to the ailing Schrödinger while Niels Bohr sat on the edge of the bed, continuing to argue.

Bohr and Heisenberg were separated from Schrödinger by basic philosophical convictions and they were unable to reach common ground from which to consider the atom. Each of them accepted and used the same body of experimental evidence, but they could not agree on the conceptual means to embrace the evidence. Schrödinger looked at the natural world and saw continuity, so he was intellectually offended by energy states and quantum jumps. As he said to Bohr, "If all this damned quantum jumping were here to stay, I should be sorry I ever got involved with quantum theory."[15]

Fortunately, Schrödinger did have something to do with quantum theory. From his Christmas vacation in Arosa through the first half of 1926, Schrödinger wrote six papers that guided and shaped both physics and chemistry in the decades that followed. In the first paper, the hydrogen atom played a decisive role. Schrödinger's treatment of the hydrogen atom found in that paper has essentially been duplicated in scores of textbooks from then to the present time. Schrödinger returned to the hydrogen atom in the third paper of the series, in which he applied his quantum mechanical approach to what is known as the Stark effect. If hydrogen atoms are placed in an electric field, the spectral lines associated with the Balmer series are altered. Specifically, each spectral line is split into a close-knit group of lines. Schrödinger successfully accounted for this splitting, called the Stark effect, and thereby used the hydrogen atom to explain it.

With his quantum mechanics, Schrödinger created a powerful formalism for treating atoms and molecules. Schrödinger's version of quantum mechanics was preferred by physicists through the 1930s. As the years passed, however, physicists mastered the matrix mathematics that was the basis of Heisenberg's quantum mechanics and with that mastery they discovered that some problems are treated more naturally with the matrix approach. Today, both approaches are used equally.

The application of quantum mechanics to physical problems is now routine with most physicists. It is used daily to guide the design of experiments and to explain their results. Every prediction made by means of quantum mechanics has been accurate. It has been an enormously successful physical theory, yet one that no physicist will claim to understand. From the beginning it was apparent that quantum mechanics required a new and novel way of thinking about the natural world and about reality. That brings us back to Schrödinger's cat.

Schrödinger was increasingly uncomfortable with the interpretation of quantum mechanics that gained widespread acceptance. This interpretation was, not surprisingly, largely influenced by Bohr and was called the Copenhagen interpretation. In the Copenhagen interpretation, if a physical system can exist in more than one state, say two states, the physical system exists in a condition that is "smeared" between the two states. In the jargon of quantum mechanics, it is said that the system exists in the superposition of two states. Schrödinger did not like this ambiguity. To convey his intellectual discomfort, he introduced his imaginary cat. Hypothetically, the cat was enclosed in a chamber with a radioactive substance selected so that in a certain time period there is a 50–50 chance of decay. If decay occurs, the cat dies; if decay does not occur, the cat lives. So the cat, in terms of the Copenhagen interpretation of quantum mechanics, exists simultaneously in two states. It is neither alive nor dead, but half-alive and half-dead—clearly, a paradox.

The cat paradox was presented in a paper, "The Present Situation in Quantum Mechanics," which Schrödinger wrote in 1935.[16] Since then, Schrödinger's imaginary cat has been a provocative source of debate. Of course, the paradoxical nature of Schrödinger's cat is, at least in part, couched in the act of applying concepts designed to account for the behavior of a subatomic object like an electron to a macroscopic object like a cat. This situation, however, is changing: the cat appears to be emerging from the murky realm of the imagination. A very ingenious experiment, designed by David Wineland and his associates, has been designed for a system intermediate between the subatomic and macroscopic domains. It promises to shed light on the shadowy boundary between the world of the electron and the world of human-sized objects.[17] This experiment is a close analog of the imaginary cat in its imaginary chamber. The results are still being debated, but it seems that the paradoxical situation continues to exist on the small scale of this experiment. Further experimentation and careful analysis are needed. It may be that physicists have not yet created an atomic version of a fully creditable cat, but it does appear that they may have produced quantum-like kittens.

It would be a stretch to link the hydrogen atom with Schrödinger's cat, yet the hydrogen atom has played an integral role in the development of quantum mechanics. Quantum mechanics has brought physicists face to face with intellectual challenges that defy resolution. It is not surprising that the enigmatic cat is a part of this conundrum.

9

The Hydrogen Atom and Dirac's Theory of the Electron

Paul Dirac, 1928

> With relativity and quantum mechanics, antimatter was part of the general framework . . . of modern physics.
>
> —Maurice Jacob

The reputation of a physicist can be built upon a body of work so important that it defines a field of research and influences all other investigators in the field. Or a physicist may write one paper that is so prominent in a domain of research that the identity of the physicist becomes linked to that epochal paper. Although the British physicist P. A. M. Dirac wrote many important papers, starting with his first paper on quantum mechanics in 1926, it is his paper on the electron, published in 1928, that comes to the minds of nearly all physicists when Dirac's name is mentioned. This paper, "The Quantum Theory of the Electron, I" is one of the great papers of twentieth-century physics.

It begins: "The new quantum mechanics, when applied to the problem of the structure of the atom with point charge electrons, does not give results in agreement with experiment. This discrepancy consists of the 'duplexity' phenomena, the observed number of stationary states for an electron in an atom being twice the number given by the theory."[1]

The theory Dirac refers to is quantum mechanics, which did in fact provide the formal and conceptual frameworks for both the analysis and synthesis of atomic structure. However, when quantum mechanics was created by Heisenberg and Schrödinger in 1925, one basic property of the electron, discovered in October, was certainly unknown to Heisenberg and probably to Schrödinger. This property of the electron is its inherent angular momentum, or spin, which makes it behave like a tiny bar magnet possessing a magnetic moment.

The electron's spin exhibits itself when the electron is in a magnetic field as the spin, or magnetic moment, orients itself either in the direction of the magnetic field (parallel to the field) or in the opposite direction (antiparallel). The energies associated with these two alignments, parallel and antiparallel, are very slightly different; thus, the electron's spin has the effect of splitting a single energy state into two slightly different energy states, and two spectral lines are often observed instead of a single line. Since it requires fairly good spectroscopic equipment to observe the effect of the electron's spin, it escaped notice early on. In any event, this is the "duplexity phenomena" referred to by Dirac.

As soon as quantum mechanics was developed, physicists faced the challenge of integrating quantum mechanics and relativity theory. Heisenberg and Pauli, along with others, tried to achieve this unification, but their approach was reminiscent of Sommerfeld's effort to bring relativity into Bohr's model; namely, relativity was treated as an add-on, grafted onto quantum mechanics rather than springing from the roots of the two theories. Dirac, who had been fascinated with the theory of relativity from his youth, had something different in mind. Dirac wanted a synthesis of relativity and quantum mechanics founded on the general principles of both theories. Dirac attended the Solvay Conference in Brussels in October 1927 and upon his return to Cambridge, he focused his attention on bringing the two great physical theories together.

Paul Dirac was a unusual person. Perhaps because Dirac's father demanded that his young son use French rather than his native English to converse with him, the young Dirac adopted the habit of silence during his childhood simply because he could not express his thoughts in French. Whatever the reason, the adult Paul Dirac was a man of silence. Dirac's silence was so intense that it inspired a little levity among physicists. In physics, the units given to physical quantities like time or length are important. Physicists, clearly in jest, have defined the unit of silence as the dirac.

Stories about Dirac are legendary. In one, when a comment was made after a Dirac talk, Dirac said nothing. Finally, after the silence continued to unbearable lengths, the host asked Dirac if he had a response. Dirac's reply was that no question had been asked. He was exacting in his response to words; he was also precise in his use of words. He once said to Bohr that one should not start a sentence unless one knows how it will end. He crafted his sentences with such care that when he was asked a question, he would often repeat the same sentences in the exact same words he had used earlier (see Figure 9.1).

Dirac's penchant for silence had an accompanying effect. He worked alone. After he was appointed Lucasian Professor at Cambridge in 1932, the position once held by Sir Isaac Newton, he did not accept many research students. His colleagues knew very little about what Dirac was doing until his work appeared in completed form. Sir Nevill Mott, a Cambridge colleague who was about as close to Dirac as anyone, said, "all Dirac's discoveries just sort of fell on me and there they were. I never heard him talk about them, or he hadn't been in the place chatting about them. They just came out of the sky."[2]

Dirac was drawn to basic questions. In addition to his need for solitude, Dirac's consuming interest in fundamental questions may have influenced his reluctance to accept students, who typically lack the maturity to grapple with nagging issues at the foun-

Figure 9.1 Paul Dirac.

dational level. He was not inclined to spend time applying physical theories to solve problems, even challenging and interesting problems. Rather, Dirac was interested in fundamental laws at work in the natural world. He believed that beauty in the mathematical expression of a theory was an indication of its validity. This belief was so important that he would not be discouraged if his ideas did not agree with experimental results. In his own words:

It seems to be one of the fundamental features of nature that fundamental physical laws are described in terms of a mathematical theory of great beauty and power, needing quite a high standard of mathematics for one to understand it. . . . Just by studying mathematics we can hope to make a guess at the kind of mathematics that will come into the physics of the future. . . . It may well be that the next advance in physics will come along these lines: people first discovering the equations and then needing a few years of development in order to find the physical ideas behind the equations.[3]

With mathematical beauty as a guiding principle, Dirac began his effort to bring the theory of relativity and quantum mechanics together in October 1927. By Christmas, he had succeeded.

Dirac imposed severe demands on his work as he set out on his quest for a wave equation for the electron that satisfied the demands of both relativity theory and quantum mechanics. He wanted no arbitrary assumptions to clutter the logical path from his starting point to his end result. Of course, there were intellectual struggles that Dirac had to work through. One particular struggle occurred when Dirac attempted to use Pauli's formal description of electron spin. The principal breakthrough came only when Dirac realized that Pauli's formalism was imposing an unneeded restriction on his thinking. When he relaxed this limitation, he arrived rather quickly at the desired result—a wave equation for the free electron. His paper, "The Quantum Theory of the Electron," was received by the editors of the *Proceedings of the Royal Society* on January 2, 1928 and was published one month later.

Often the work of a first-rate scientist contains more than intended, more than planned, more than anticipated. This was the case with Dirac's theory of the electron: unexpected results fell out naturally from his theory. One result of Dirac's work was

certainly planned—a result that would have doomed the theory
had it not been achieved. Coming out of Dirac's theory, as a natural consequence and with no arbitrary assumptions, was the fine-
structure formula that Sommerfeld had derived thirteen years
earlier. However, Dirac's theory did much more: it gave the correct energy states for the hydrogen atom complete with the spectral fine structure that results from the spin of the electron. As
Helge Kragh, Dirac's biographer, states, the spectrum of the hydrogen atom was "a showpiece for the Dirac equation."[4] We shall
see in Chapters 15 and 16 that the hydrogen spectrum still presented challenges, but in 1928, the Dirac equation passed a crucial
test with its successful treatment of the hydrogen atom (see Figure 5.3).

Dirac's theory of the electron accomplished even more than its
success with describing hydrogen. Dirac's approach to the electron was from first principles; he did not introduce the electron's
spin in advance. Yet, the correct spin of the electron came out as a
natural consequence of Dirac's results. Just as his theory of the
electron accounted for the spectrum of hydrogen, so it accounted
for the spin of the electron. The electron's magnetic property, its
spin, was given in terms of the electron's charge and its mass.
Thus, from the general principles of relativity and quantum mechanics, spin was understood. Dirac stated: "I was not interested
in bringing the spin of the electron into the wave equation, did
not consider the question at all. . . . The reason for this is that my
dominating interest was to get a relativistic theory [of quantum
mechanics]. . . . It was a great surprise for me when I later discovered that the simplest possible case did involve spin."[5] When the
physicists at Göttingen learned of Dirac's achievement, they were
astounded. Léon Rosenfeld was working with Max Born when the
news of Dirac's results arrived. The deduction of spin "was regarded as a miracle," said Rosenfeld. "The general feeling was
that Dirac had had more than he deserved! Doing physics in that

way was not done! . . . It [the Dirac equation] was immediately seen as *the* solution. It was regarded really as an absolute wonder."[6] On February 13, 1928, Heisenberg wrote to Dirac, "I admire your last work about the spin in the highest degree."[7]

Another very important and totally unexpected result came from Dirac's integration of relativity and quantum mechanics. This result was not only unexpected, it was so troublesome that in spite of the awe Dirac's paper generated, physicists' response was tempered by this bizarre result. The result arose because Dirac's equation accounted for ordinary, negatively charged electrons, but the solution of the equation also inferred some kind of positively charged particles. These "positive particles" were given various interpretations over the three years following publication of Dirac's paper. One interpretation advanced by Dirac himself was that the "positive particles" were holes among electrons. Such holes, he reasoned, would appear to be positive in "the sea of negative electrons." Later, Dirac identified the positive particles with protons, but J. Robert Oppenheimer showed that such an interpretation was untenable. Finally, in May of 1931, Dirac introduced the idea that the positive particle coming out of his equation was a positively charged electron—an antielectron. In a paper published in 1931, Dirac wrote, "A hole, if there were one, would be a new kind of particle, unknown to experimental physics, having the same mass and opposite charge of the electron."[8] During the following year, Dirac's "new kind of particle" was discovered by Carl Anderson at the California Institute of Technology. The new particle, which came to be called the positron, was the first antiparticle discovered. More significant, the discovery of the positron brought recognition that antimatter existed more generally.

Dirac did not actually predict the existence of the antielectron. He did not call for physicists to set up experiments designed to discover the positron. It can be argued that it would have been out of character for Dirac to give voice to a prediction. True to his character, he was silent and quietly waited. Many physicists would have called attention to their work by making a prediction, but Dirac was not concerned about the actual existence of the particle that was suggested by his theory. His sense of fulfillment came from the beauty of the theory itself.

Thanks to Dirac's electron, it is now recognized that all the particles that make up the universe have antiparticles. The existence of antimatter was another unanticipated result that fell out of Dirac's prolific equation. It is a wonder that mathematics can be used so powerfully to express the content of nature's laws. Mathematics is a product of mind and is independent of the outside world; thus, mathematicians can create great new mathematical systems without any thought given to planets revolving around stars, light passing through a prism, atoms combining into molecules, molecules working cooperatively in an amoeba, neurons firing in the brain of a youngster, or a physicist reflecting on why mathematics works as it does to express the laws of nature so efficiently. Sometimes the mathematics is quite simple; other times it can be very abstruse. This mysterious relationship between mathematics and physics prompted James Jeans to suggest that the Great Architect of the universe is a mathematician.

Dirac's mathematical equation can be written in one line:

$$\frac{ih}{2\pi}\frac{\partial|\psi\rangle}{\partial t} = (c\vec{u}\cdot\vec{P} + \beta mc^2)|\psi\rangle.$$

From the solution of this little equation come details about the hydrogen atom, the spin of the electron, and the existence of antimatter. Poets bring us fresh insights with the right sequence of words; Dirac brought us fresh insights with the right sequence of symbols.

According to Dirac's wife, Margit, there were few things that

Dirac dreaded more than reporters. He avoided them whenever possible. He even managed to avoid the press in 1933, one year after the discovery of the positron, when he and Schrödinger were awarded the Nobel Prize. Dirac's prize was given in recognition for his role in the creation and development of quantum mechanics. In announcing the prize, a London newspaper described Dirac "as shy as a gazelle and modest as a Victorian maid."[9] The London reporter said it just right.

10

Hydrogen Guides Nuclear Physicists:
The Discovery of Deuterium

Harold Urey, 1932

> Because of its unique properties, the distribution of deuterium in the universe constitutes a powerful clue to the history of the development of matter.
>
> —David N. Schramm and Robert V. Wagoner

Nearly all the chemical elements that make up our material world occur in different isotopic forms. Every breath we inhale brings into the lungs oxygen in three isotopic forms: O^{16}, O^{17}, and O^{18}. Each of these oxygen atoms has eight electrons around a nucleus with eight protons. In addition, O^{16} has eight neutrons, O^{17} has nine, and O^{18} has ten neutrons in their nuclei. The eight electrons and protons give oxygen its place in the Periodic Table of the Elements, and the eight electrons also determine oxygen's chemical behavior. Thus, the chemical behavior of each isotope of oxygen is essentially the same. However, the masses of the oxygen isotopes differ because of the different number of neutrons. This mass inequity, as we shall see, influences the physical behaviors of the three isotopic forms of oxygen.

The discovery of isotopes was made by Frederick Soddy in 1910. Earlier, Soddy had worked with Ernest Rutherford at McGill University, where they laid out the basis for understanding radioactivity in their paper "The Cause and Nature of Radioactivity."[1] From Montreal, Soddy went first to London for one year and then on to Glasgow University, where he discovered isotopes—a term he coined in 1913.[2] Soddy and Rutherford showed that in the process of radioactive transformations, the atoms of one chemical element could be transmuted into the atoms of another chemical element. In addition, it became apparent to Soddy that a radioactive transformation could give rise to atoms that differed in their weights but were chemically identical. Such atoms are isotopes of the same chemical element.

Soddy's discovery of isotopes put Prout's provocative idea that hydrogen was the basic building block of the chemical elements into the inactive archives of science history. The observed weights of the chemical elements are determined by their isotopic composition, which failed to be multiples of hydrogen, as Prout believed it would be. The atomic weight of oxygen, for example, is 15.9994, and that of chlorine is 35.453.

Deuterium is an isotope of hydrogen, often called heavy hydrogen. The most abundant isotope of hydrogen is H^1, with a single electron in orbit around a single proton. In deuterium, H^2, a neutron joins the proton in the nucleus. A naturally occurring sample of hydrogen consists of 99.985 percent H^1 and 0.015 percent H^2; thus, ordinary hydrogen (H^1) is abundant whereas heavy hydrogen (H^2) is scarce.[3]

The discovery of deuterium is, according to one of the physicists who participated in the experiments that led to its discovery, a "story of missed opportunities and errors."[4] In 1913, two scientists at New York University held the discovery of deuterium in their hands when they measured the density of water to great accuracy. They found that the density of water varied from sample to sample and concluded that pure water does not possess a unique density. Their results varied according to the varying presence of deuterium in their samples. This was the first experimental evidence that provided a hint of the existence of deuterium. Had these scientists responded to their results with an experiment to distill water into fractions with different molecular weights, they may well have discovered deuterium twenty years early, but the discovery slipped through their fingers.

In 1929, Harold Urey and Berkeley chemist Joel Hildebrand left their hotel, hopped a taxi, and headed to their conference. Ferdinand G. Brickwedde was with them and listened to their conversation. Hildebrand informed Urey that chemists at Berkeley had just discovered the isotopes of oxygen, O^{17} and O^{18}. Hildebrand said, "They could not have found isotopes in a more important element." Urey responded, "No, not unless it was hydrogen."[5] In 1931, just before Urey started his own experiment to determine whether hydrogen had an isotope, two physicists at Berkeley were examining the physical and chemical bases for establishing atomic weights. The two approaches led to slightly different results. From this work they concluded that hydrogen was a mixture of isotopes—mostly H^1 and a small amount of a heavier form. This work was reported in the July 1, 1931 issue of *Physical Review*. When Urey received this journal, he immediately began planning his investigation.

Urey's experiment had two parts. First, since a sample of hydrogen contains approximately one atom of deuterium, H^2, for every 7,000 atoms of hydrogen, H^1, a method to increase the concentration of the heavier isotope was necessary. A sample somehow enriched with deuterium would make it easier to detect deuterium's feeble presence and the results would be more definitive. Second, an experimental method to detect the presence of deuterium was needed. The experimental plan was: concentrate and detect.

Urey first tried to detect deuterium directly by using a sample of bottled hydrogen. Urey, a professor of chemistry at Columbia University, brought his colleague George Murphy on board and together they set up the apparatus for a careful spectroscopic study of hydrogen. Specifically, they designed their experiment to produce the spectral lines of the Balmer series. The two atoms of hydrogen, light and heavy, would give rise to spectra that were essentially the same, except the wavelengths of the spectral lines associated with the heavy isotope would be slightly shifted relative to the wavelengths of the lighter isotope. They did see very faint lines at the wavelength positions their calculations suggested. Thus, they believed they were observing evidence for the heavy form of hydrogen. They wanted stronger evidence, however. They did not want to be misled by possible impurities or some instrumental error. Urey then decided that a method must be found to increase the concentration of deuterium in the sample to be analyzed.

For this challenge, Urey went to the National Bureau of Standards, where he talked to Brickwedde. Urey's idea was to distill liquid hydrogen and concentrate the heavier form of hydrogen. This method of concentrating deuterium relies on a physical behavioral difference arising from the disparity between the masses of the two isotopes. At temperatures below 20.4 degrees kelvin (K), or $-252.6°C$, hydrogen is a liquid in the form of molecular hydrogen, H–H or H_2. Most of the molecules consist of two ordinary hydrogen atoms: H–H. A small fraction of the molecules bring together the ordinary and heavy forms of hydrogen: H–D. In the liquid, the H–H molecular form moves around a little faster than H–D because it is less massive. As the liquid is slowly evaporated, the faster-moving H–H is more likely to leave the surface of the liquid; H–D molecules are more likely to stay behind, hence the liquid becomes slightly more concentrated with H–D. Urey's plan was to evaporate some 6,000 cm^3 (six liters) of liquid hydrogen down to a volume of about 2 cm^3, which, according to plan, would be greatly enriched in the heavier isotope of hydrogen.

This was Urey's proposal to Brickwedde. Brickwedde agreed to help. In his first attempt, Brickwedde evaporated liquid hydrogen at a temperature of 20K. But some procedural errors negated the

intended outcome and Urey detected no enhanced presence of
deuterium in the Balmer spectral lines. In his next try, Brickwedde
evaporated liquid hydrogen at a lower temperature of 14K. The
sample of hydrogen resulting from this distillation was indeed
richer in the heavy isotope, and the Balmer spectral lines corresponding to deuterium were more intense by a factor of six or
seven. On the basis of these results, Urey concluded that the isotope of hydrogen, deuterium, really existed.

Papers reporting the discovery were published in early 1932. In
1934, Urey won the Nobel Prize in chemistry for the discovery of
deuterium.

Harold Urey was an unusual scientist. He grew up in Montana,
the son of poor, homesteading parents. After graduating from
high school, Urey taught for three years in a small country public
school. He attended Montana State University, studied zoology
and chemistry, and graduated in 1917. To pay for his education,
Urey worked summers on a railroad road gang laying track in the
Northwest. During the academic year, Urey lived in a tent.

Soon after his discovery of deuterium, Urey's exceptional character revealed itself in a very selfless way. In honor of his discovery, Urey received a prize from the Carnegie Foundation.
The prize was for $7,600—a large amount of money in the early
1930s. In an act of rare generosity, Urey gave half the money to a
young and, in his judgment, promising physicist colleague, I. I.
Rabi. As Rabi later recalled: "Urey did one of the most extraordinary things imaginable. He gave me half of it [the money]. I had
nothing to do with the discovery. What a greatness in Harold
Urey—what a tremendous magnanimity to do something like
that."[6] The money from Urey allowed Rabi to improve and expand his molecular beam experiments. As we shall see in Chapters
11, 12, and 13, Urey's judgment about Rabi's potential was well
founded.

It is unusual for a Nobel Prize to be awarded so quickly after a discovery. But because of its simplicity, hydrogen attracts attention. Urey's discovery stimulated an outburst of research activity. Within two years of its discovery, over one hundred papers were published that involved the new isotope of hydrogen. In 1934, another one hundred deuterium-related papers were published. Although both chemists and physicists were intrigued by this new isotope of hydrogen, it was the nuclear physicists who began a long affair with deuterium.

The nucleus of deuterium is called the deuteron. In a basic sense, the deuteron is to physicists interested in the nucleus what the hydrogen atom is to those interested in the atom. The hydrogen atom is the simplest atom; the deuteron is the simplest compound nucleus. Of course, the nucleus of the hydrogen atom is the simplest nucleus of all, but it is a lone proton and does not bring into play any of the forces that hold larger nuclei together. The deuteron, one proton and one neutron bound together, is the nucleus of the deuterium atom. The question that fascinates nuclear physicists is this: What is the nature of the force that binds the proton and neutron together?

The parallels between the hydrogen atom and the deuteron are provocative. Both are two-particle systems: the hydrogen atom, an electron and a proton; the deuteron, a proton and a neutron. Both can be treated in similar ways. There is, however, one significant difference. Whereas the force between the electron and the proton in the hydrogen atom is the well-known Coulomb force, the force that acts between the proton and neutron inside the deuteron was not known in the early decades of the twentieth century. The need to understand the nature of the force between nuclear particles has made the simple deuteron the playground of nuclear physicists. In fact, the deuteron has been studied more than any other nucleus, and the insights gained from this simple nucleus have guided physicists as they grapple with more and more complicated nuclei. As we shall see in Chapter 14, the deuteron eventually forced a complete overhaul in the thinking about nuclear forces.

Once again, the hydrogen atom was a source of inspiration—this time for a chemist. The discovery of deuterium by Harold Urey has guided the thinking and experiments of physicists as they have sought to expose the forces at play inside the atomic nucleus.

Naming the newly discovered isotope of hydrogen was far more time-consuming, as it turned out, than Urey's experiments themselves. Normally, the honor of naming the newly discovered isotope would go to the discoverer, Harold Urey. But many leading physicists, including Earnest Lawrence, G. N. Lewis, Ernest Rutherford, R. A. Millikan, and others (including professors of Greek, one from Columbia and the other from Berkeley) joined in the controversy. With all their intelligence, it took these individuals and the larger community two years to agree on the name deuterium—which was Urey's choice.[7]

The deuterium story is revealing. Physics, based on hard data, is typically easy for scientists to agree on. But deciding on the name of something brings out emotions and vested interests that can spark disagreements and challenge friendships. Discovering deuterium was relatively straightforward; naming it was another matter.

11

Hubris Meets Hydrogen: The Magnetic Moment of the Proton

Otto Stern, 1933

When Otto Stern measured the proton moment in the early 1930s, he was advised not to bother—elementary theory proved the result would be one nuclear magneton. Fortunately, Stern had a healthy disregard for elementary theory.

—Daniel Kleppner

The name of Otto Stern is identified with one of the most bizarre and influential experiments in quantum physics. The Stern-Gerlach experiment was carried out prior to the creation of quantum mechanics and for some physicists, this experiment, more than any other, demonstrated convincingly that the physics of the nineteenth century, powerful though it was, could not describe the strange behavior of atoms. Up to 1922, when Stern and Walther Gerlach did their experiment, many physicists courted the hope that the quantum ideas contained in the works of physicists like Max Planck, Albert Einstein, Niels Bohr, and Arnold Sommerfeld would prove to be a passing fancy. Stern himself was one of these physicists. In response to Bohr's quantum model of the hydrogen atom, Stern and his friend Max von Laue, also an eminent physicist, agreed that if the crazy model of Bohr was correct, they would quit physics.[1]

The Stern-Gerlach experiment did not involve hydrogen directly, but did so indirectly. The epoch-making Stern-Gerlach experiment demonstrated the physical reality of space quantization that Sommerfeld had proposed as an outgrowth of his study of the hydrogen atom. In this study, Sommerfeld maintained that when a magnetic field is applied to an atom, its electron orbits could take only certain orientations in space. Just as iron filings are observed to align themselves along the magnetic field lines, so in an analogous fashion electron orbits align themselves in space. More specifically, Sommerfeld concluded that only certain orientations relative to the spatial direction of the applied magnetic field were allowed. Sommerfeld postulated space quantization to explain the splitting of hydrogen's spectral lines in the presence of the magnetic field. This splitting is called the Zeeman effect after Pieter Zeeman, who discovered the splitting effect of a magnetic field in 1896.

Bohr applied a quantum condition to the energy states of the hydrogen atom: only certain energy states were allowed. Sommerfeld applied a quantum condition to the orientation of electron orbits: only certain spatial orientations relative to an applied magnetic field were allowed. The experiment of Stern and Gerlach was designed to test Sommerfeld's explanation of the Zeeman effect, namely, the idea of space quantization.

Stern's experiment was conceptually simple and, as such, it had a beauty all its own. His approach was based on the method of molecular beams. The molecular beam method was originated in 1911 by Louis Dunoyer. In 1921, it was a relatively novel experimental method. Since that time, the molecular beam method has been the basis for an extremely productive line of physical investigation and, as we shall see in future chapters, has yielded both detailed and precise information about atomic properties. In this method, atoms diffuse from a source at one end of a highly evacuated cylindrical chamber, travel a path along the axis of the chamber, and are detected at the other end. Near the source exit, a succession of narrow slits collimate, or make parallel, the atoms into a narrow, ribbonlike band. The rate of diffusion from the source is slow so that there are only a small number of atoms in the beam; thus, individual beam atoms are widely separated and each atom moves in quiet isolation along the path through the vacuum of the chamber. This is the basis for the beauty of the molecular beam method—it combines simplicity and power. Each atom in the beam is isolated so that it is free of all external influences from neighboring atoms, and along the path traversed by the beam atoms, various influences can be designed to exert precisely known forces on each individual atom. The atom's response to these predetermined influences can be measured by a detector at the end of the chamber.

In the Stern-Gerlach experiment, it was a magnetic field that interacted with the passing atoms. When the atoms struck the detector, a very cold glass plate, they left a record as atom after atom deposited itself on the cold plate (see Figure 11.1). Some physicists thought Stern's plan to test the idea of space quantization was rather silly. Many physicists did not regard space quantization as "real"; rather, they regarded it as merely a calculational device hatched by Sommerfeld to explain a particular set of data. That is exactly what Peter Debye thought. He said to Stern, "But you surely don't believe that the [spatial] orientation of atoms is something physically real; that is [only] a prescription for the calculation, a timetable for the electrons."[2] Not deterred, Stern persevered.

Stern used a beam of silver atoms for his experiment. As the silver atoms streamed toward the detector, they were nudged, at a right angle to their motion, by an applied magnetic field that caused the beam to spread out. Thus, the detector "saw" a slightly widened beam of silver atoms. The doubters, like Debye, expected the beam atoms to be distributed *continuously* across the widened dimensions. Stern and Gerlach found something dra-

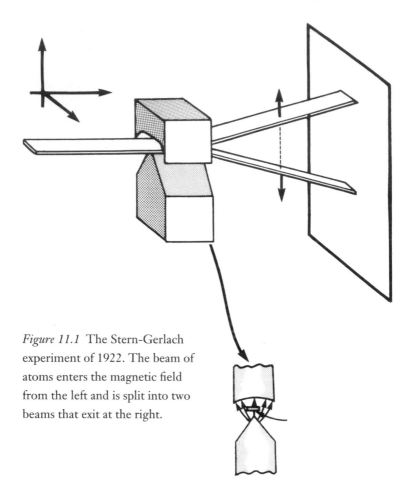

Figure 11.1 The Stern-Gerlach
experiment of 1922. The beam of
atoms enters the magnetic field
from the left and is split into two
beams that exit at the right.

matically different: the beam atoms were concentrated in two,
slightly separated regions. Two faint fringes were deposited on
the detector. The silver atoms were obviously constrained: they
were not free to pick their destination. It was one fringe area or
the other. There was no way to explain the result except to assume
space quantization. It was a difficult experiment, but in late 1921
Stern and Gerlach produced incontrovertible evidence that space

quantization is real. With this result, however, nature played a joke on Stern and other physicists: while Stern's results confirmed space quantization, it was not the orbits that oriented themselves, as Sommerfeld had proposed. Ironically, the space quantization idea itself was correct, but it was caused by a property of the electron unknown in 1922—electron spin. It was the electron's spin, or its magnetic moment, that was orienting itself, not the electronic orbits.

The conceptual simplicity of the Stern-Gerlach experiment, coupled with the directness of its results, provided commanding evidence for the quantum theory. I. I. Rabi was a graduate student at Columbia University when the Stern results were announced. The Stern experiment changed forever Rabi's thinking about quantum mechanics. "This convinced me once and for all," Rabi said later, "that an ingenious classical mechanics was out and that we had to face the fact that the quantum phenomena required a completely new orientation."[3]

Shortly after the results of the Stern-Gerlach experiment appeared in the scientific literature, Stern received an invitation to join the faculty at the University of Hamburg, where, over the period 1922 to 1933, he continued his experimental work. In 1932, Stern decided to adapt the molecular beam method to a daunting experiment: to measure the magnetic moment of the proton, the nucleus of the hydrogen atom. Joining him is this endeavor was Otto Robert Frisch.

By 1932, spin had come to be recognized as a basic property of electrons and protons. These charged particles behave as though they spin around an axis through their center. This spinning charge gives rise to a magnetic moment that makes electrons and protons behave like a tiny bar magnet. In an external magnetic field, these magnetic moments, or spins, align themselves in ways correctly described by quantum mechanics. In atoms, the spins of electrons and protons typically pair up and cancel each other's

magnetic moment. In some atoms, however, there is an odd number of electrons so that the atom has a net magnetic moment.
This is the case with the silver atom. In his 1922 experiment with
Gerlach, Stern observed the alignment of silver's magnetic moment, either with the field or against it, leading to the two discrete
fringes observed at the detector.

The hydrogen nucleus consists of a single proton and it has a
resulting magnetic moment. This is what Stern wanted to measure. For several reasons, this measurement promised to be exceedingly difficult. To appreciate these difficulties, let us momen-
tarily return to Dirac's famous work on the electron. The fruitful
Dirac equation, among other results, provided directly the size of
the electron's magnetic moment. This was one of the triumphs of
Dirac's theory. His electron result was so convincing that it led
physicists to assume that they already knew the magnetic moment
of the proton. They were so confident that they derided Stern's
willingness to take on such a difficult and meaningless experiment.

Extending Dirac's theory, physicists assumed that the magnetic
moment of the proton was smaller than the electron moment; in
fact, they assumed it was exactly $1/1,836$ times smaller. This exact
number comes from the mass of the proton, which is 1,836 times
larger than the mass of the electron. This was the first difficulty
Stern faced: the tiny size of the proton property he resolved to
measure. In the hydrogen atom, the proton is coupled with the
electron and it was clear to Stern that the electron's magnetic moment, 1,836 times larger, would overwhelm the smaller effects of
the proton. In response to this, Stern decided to use two hydrogen atoms chemically bonded to each other in the form of molecular hydrogen. In the hydrogen molecule, the magnetic moments
of the two electrons are oriented oppositely and thus cancel each
other out. With this difficulty resolved, Stern faced other difficulties.

As a molecule moves from point to point, it also tumbles end-over-end. This rotational tumbling of the molecule's electrons and protons gives rise to a rotational magnetic moment that Stern would have to separate out from the magnetic moment of the proton. Stern was able to devise a way to determine these rotational effects. Every hydrogen molecule brings together two hydrogen atoms, each with its own proton nucleus. The magnetic moments of these two protons can be aligned either parallel or antiparallel to each other. In Stern's experiment, these two alignments would respond differently to the subtle pushes and pulls each molecule received as it traversed the path through the magnetic fields of the molecular beam apparatus; in fact, the magnetic fields exert no force on the proton spins in those hydrogen molecules with anti-parallel alignments. Thus, Stern used the deflection of the hydrogen molecules with antiparallel moments to determine the rotational magnetic moment and then, with the rotational magnetic moment known, he could account for the rotational contribution to his final data. The final challenge Stern faced was that the molecular beam method would have to be pushed to its limits to detect the minute magnetic moment of the proton. There were no comfort margins, no margins for sloppiness, no margins for error. Everything would have to work just right and do so simultaneously.

When he began the experiment, Stern made the same fashionable assumption that all his colleagues made; namely, he assumed the magnetic moment would be 1/1,836 the size of the electron's magnetic moment. As the experimental data began to accumulate, however, it became clear that something unexpected was occurring. The proton's magnetic moment was looking larger than the predicted value. Of course, Stern examined every aspect of the experiment and examined all the assumptions he brought into his analysis. The data were rough and Stern was not convinced.

About this time, Stern gave a seminar on the subject of his experiment. He asked the audience to take out a piece of paper and write their prediction of the outcome: What would be the magnitude of the proton's magnetic moment? He further asked them to sign their prediction. With all the confidence of a weather man predicting rain while standing in a downpour, the assembled physicists all predicted the catechistic answer learned from Dirac. The great Pauli told Stern, "If you enjoy doing difficult experiments, you can do them, but it is a waste of time and effort because the result is already known."[4] Stern collected the slips of paper with each physicist's prediction and calmly puffed on his cigar. Even with the rather imprecise data, Stern was confident that a surprise awaited his colleagues. To confirm his confidence, more and better experimental data were needed.

For the second round of experiments, a new student, Immanuel Estermann, worked with Stern. Stern's student assistants were important. Like many experimental scientists, Stern was not particularly good with his hands. He was supreme at conceptualizing experiments. He was great at interpreting experimental results. But his students shuddered when Otto Stern touched the apparatus. Estermann was adept with his hands.

On the basis of the results of the first experiments, Stern and Estermann redesigned the apparatus. They now assumed that the magnetic moment was larger than earlier assumed and this allowed them to relax the strain on the apparatus and improve the precision of their measurements. They did the experiment in two different ways. The result from both methods gave the same results. The magnetic moment of the proton was about 2.5 times greater than predicted.

The measurement of the magnetic moment of the proton was not very precise. Stern claimed that his result had an uncertainty of 10 percent. This means that his result could range from 2.25 to 2.75 nuclear magnetons. Such a range of uncertainty was not acceptable to Stern. He was tooling up to remeasure not only the proton's magnetic moment, but to measure the deuteron's magnetic moment as well. Harold Urey's discovery of deuterium had just been announced.

Stern's plan was compromised as a black cloud spread itself over his experimental work in Hamburg. On January 30, 1933, Hitler became chancellor of Germany and Estermann, a Jew, was notified that his appointment at the University of Hamburg would be terminated. Stern was also a Jew, but because he was a German veteran of World War I, he was briefly exempt from Hitler's edict. Nonetheless, Stern did not wait: he immediately resigned his position in protest. Thus, the final experiments in Hamburg on the proton and the deuteron were carried out under very trying circumstances. His result for the deuteron was very tentative and largely qualitative. In the summer of 1933, Stern's molecular beam laboratory in Hamburg closed up shop. Stern and Estermann were both invited to come to the United States, and Stern joined the faculty of the Carnegie Institute of Technology, where they continued their work on the proton and the deuteron. Frisch, also a Jew, left for London. In 1943, Otto Stern won the Nobel Prize for physics for his measurement of the proton's magnetic moment.

The unit used to express the magnetic moment of the proton is the nuclear magneton. This clumsy unit is patterned after the unit for the electron magnetic moment which is called the Bohr magneton. The magnetic moment of the electron is 1 Bohr magneton. The *predicted* magnetic moment for the proton, based on an extrapolation from Dirac's 1928 paper, was 1 nuclear magneton. Stern's *measured* result was 2.5 nuclear magnetons. Theoretical physicists badly missed the mark—they were off by 150 percent. This result demonstrated convincingly that physicists did not understand the proton.

Physicists were so impressed with Dirac's theory of the electron that they jumped to the conclusion that the same theory would apply to the proton. Their confidence and arrogance became hubris. It was the nucleus of the hydrogen atom, the proton, that confronted this hubris. Like Prometheus, chained to a rock by Zeus, physicists were intellectually chained to the rock of their erroneous prediction. Fortunately, unlike Prometheus, an eagle did not feast daily on their livers. Stern's measurement of the proton's magnetic moment using the simple hydrogen atom transformed hubris into humility.

12

The Magnetic Resonance Method:
The Origin of Magnetic Resonance Imaging

I. I. Rabi, 1938

You felt if you were measuring the properties of hydrogen, the most fundamental nucleus, you just measure it and do it as well as you can.

—I. I. Rabi

When Isidor Isaac Rabi was a graduate student at Columbia University in the mid-1920s, his eye was on Europe, where intellectual fireworks were illuminating the minds of physicists, animating their discussions, and entertaining their ambitions. This was not the situation at Columbia University, where only a sputtering roman candle occasionally lit the corridors and laboratories of the physics building, Pupin Hall. In 1926, the contrast between European and American physics was like that of the grand finale of a major pyrotechnic display and a simple sparkler.

By the time Rabi completed his doctoral work in 1926, he was eager to witness firsthand the heat and glare that occurs when new physics bursts open fresh fundamental insights into the nature of the material world. One of the many young American physicists who went to Europe in the 1920s to learn the new physics from the creators, he wanted to join in the fun. In what was typical Rabi fashion, he didn't plan ahead, he just went. He visited Erwin Schrödinger in Zurich, Niels Bohr in Copenhagen, Wolfgang Pauli in Hamburg, and Werner Heisenberg in Leipzig. While in

Hamburg, he also met Otto Stern, and he followed Stern's work with interest.

On one visit to Stern's molecular beam laboratory Rabi made a casual suggestion to Stern for an experiment, which brought the immediate response: "Why don't you do it?" Rabi was told it was an honor to receive such an invitation from Stern. "I was in no position to refuse an honor," said Rabi.[1] Rabi's experiment introduced a novel configuration of the magnetic field for deflecting particles in a molecular beam—a configuration now called the Rabi field.

In writing about molecular beams, Otto Stern once referred to "that beauty and peculiar charm which so firmly captivates physicists working in this field."[2] Stern's statement accurately described Rabi's reaction to his experience in Hamburg. Rabi's powerful physical intuition was stimulated by images of atoms moving through the apparatus and once he was exposed to that "beauty and peculiar charm," the course of Rabi's professional life was determined (see Figure 12.1).

Rabi returned to the States to join the Columbia faculty in the fall of 1929, just before the crash preceding the Great Depression. As it happened, while Rabi was still in Europe finishing his immersion in quantum mechanics, Heisenberg left Europe for an extended visit to the United States. His first stop was New York City, where he visited Columbia University. The physics department was seeking a new faculty appointment. They sought a person who could introduce a course in quantum mechanics into their physics curriculum. Heisenberg recommended Rabi. At this time, anti-Semitism worked against the appointment of Jews to university faculties, but with Heisenberg's strong recommendation, George B. Pegram, the chair of the physics department, went against the anti-Semitic prejudice that pervaded higher education and appointed I. I. Rabi Lecturer in Physics, the lowest academic rank.

Figure 12.1 I. I. Rabi.

Throughout the 1930s Rabi and his illustrious group of students and postdoctoral assistants carried out a succession of molecular beam experiments on the hydrogens—ordinary hydrogen (H) and heavy hydrogen (D). With ever-increasing accuracy and precision, the magnetic moments of the proton and the deuteron were measured. The culmination of this effort was the discovery of the magnetic resonance method, which portended a rich dividend of applications far beyond anything Rabi or his students could have imagined. That is the nature of basic research: it contains within it unanticipated rewards that often have practical applications.

When Rabi read about Stern's surprising result for the magnetic moment of the proton, he decided it was necessary to redo the experiment to verify the unexpected result. Furthermore,

Rabi discovered he could do the experiment in a different way. Stern had used a beam of molecular hydrogen, which effectively canceled out the much larger magnetic moments of the electron and permitted the proton's small magnetic moment to be detected. Stern's experiment required strong deflecting magnetic fields that were difficult to regulate, control, and calibrate. With Gregory Breit, a New York University physicist, Rabi found a way to couple the weak magnetic moment of the proton with the strong magnetic moment of the electron so that the latter effectively leveraged the former. This approach enabled Rabi to employ a beam of atomic hydrogen (as opposed to Stern's use of molecular hydrogen) and weak deflecting magnetic fields (as opposed to Stern's use of strong magnetic fields). Thus, Rabi's experimental procedure avoided many of the complications that Stern had wrestled with.

In the early 1930s, an outstanding postdoctoral assistant joined Rabi's research team, Jerrold Zacharias, who would eventually become Institute Professor of Physics at the Massachusetts Institute of Technology. At that time, however, Zacharias was a victim of the anti-Jewish sentiment that pervaded universities and he had only a temporary job. "I couldn't get one [a job] because I was Jewish," said Zacharias.[3] Zacharias recognized that Rabi was different. "Rabi was an unusual case," continued Zacharias, "in that Heisenberg had buffaloed Dean Pegram to hire him. And Rabi was the only Jew and was an unusual guy."[4]

Zacharias had also seen the paper in which Stern's proton result was described and he too was intrigued. Zacharias went to Rabi and said, "Rabi, I'll work with you if you'll work on atomic hydrogen. Atomic hydrogen is as complicated a beast as I am willing to get involved in."[5] Rabi accepted Zacharias's terms and in late 1933, Rabi and his associates began to build apparatus—apparatus that would be changed many times over the course of the 1930s and evolve into the magnetic resonance method.

The first hydrogen experiment in Rabi's laboratory had its strengths and weaknesses. A strength was the form of the magnetic field that deflected hydrogen atoms as they passed through the beam apparatus. Zacharias doubted Stern's results because he was skeptical about whether the strength of the strong magnetic field required in Stern's approach could be accurately calibrated. Zacharias conceived a method of producing a magnetic field whose strength could be calculated directly. This was a strength. The weakness was the way the hydrogen beam was detected. The detector was a glass plate coated with a chemical that turned from yellow to blue at the site where the beam of hydrogen atoms impacted upon it. This meant that subjective judgments about color had to be made—where the blue began against the background of yellow.

Because of the inherent weaknesses with the first experiment on the proton, the experimental results, reported in 1934, were given with the large uncertainty of 10 percent. But even with the imprecision of the results, they were provocative. Rabi's first measured result of the magnetic moment of the proton was 3.15 ± 0.31 nuclear magneton (nm), larger than Stern's result of 2.5 ± 0.25 nm. The two results did not agree: The largest value consistent with Stern's measurement was 2.75 nm whereas the smallest result consistent with Rabi's measurement was 2.84 nm. Although this difference raised questions that had to be pursued, Rabi's result suggested that the magnetic moment of the proton was approximately three times larger than the experts had predicted. In addition to the proton, Rabi reported the experimental results obtained for the deuteron: it had a magnetic moment of 0.77 ± 0.2 nm with a whopping uncertainty of 26 percent.

The results of the first round of hydrogen and deuterium experiments mandated new and better experiments. The proton and the deuteron are, respectively, nature's simplest nucleus and compound nucleus; thus, they are of enormous significance to a thorough understanding of atoms. The second series of experiments began in 1935.

In the second series of experiments on hydrogen and deuterium, two particularly irksome sources of error were eliminated. First, there was no longer the need to squint at a fuzzy blue color on the detector plate. This time a wire detector was devised whose electrical resistance changed when the beam particles struck it. This detection system took the guesswork out of interpreting experimental results and made the detection objective. Second, an additional deflecting magnet was added. This meant that the beam particles—hydrogen atoms—passed through two magnetic fields in succession designed to deflect the beam particles in opposite directions. Specifically, the second deflecting magnet was designed with the capacity to undo the deflection caused by the first deflecting magnet. It worked this way: with the strength of the first deflecting magnet set, no hydrogen atoms reached the detector. Then the strength of the second deflecting magnet was slowly increased until the beam particles were refocused onto the wire detector. From the value of the current producing the second magnetic field, the value of the proton's magnetic moment could be determined. The refocusing method, as it was called, took further uncertainties out of the experiment. Rabi loved it. "The experiments were beautiful. There are tricks you can play. . . . It had tremendous charm."[6]

Another change Rabi made to the second series of experiments allowed the sign of the magnetic moment to be determined, that is, to determine whether the magnetic moment was aligned parallel to the angular momentum (positive) or antiparallel (negative). Previously, neither Stern's nor Rabi's experiments could extract this information from their data. Knowing the sign of the deuteron's magnetic moment, made up of the proton's and neutron's moments, would allow them to deduce both the magnitude and the sign of the magnetic moment of the newly discovered neutron. Since the size of the proton's magnetic moment was three times larger than predicted, whether the magnetic moments were positive or negative was really unknown. "I had people to bet either way," said Rabi.[7] To determine the sign, a third magnetic field was added between the two deflecting fields. This field, called the T-field, effectively changed the intensity of the detected beam depending on whether the magnetic moments were positive or negative.

The results of the second experiment were announced at a January 1936 meeting of the American Physical Society in St. Louis: the magnetic moment of the proton equaled $+ 2.85 \pm 0.15$ nuclear magnetons, and the magnetic moment of the deuteron equaled $+ 0.85 \pm 0.03$ nuclear magnetons. From these data, the neutron's magnetic moment was deduced to be -2.0 nuclear magnetons.

The second round of experiments on the hydrogens not only gave the signs of the magnetic moments, they also gave the size of the moments with greater precision. The uncertainty in the proton's magnetic moment was reduced from 10 percent to 5 percent and for the deuteron, the uncertainty was reduced from 26 percent to 4 percent, which were considerable improvements.

In the meantime, Stern and Estermann were working in Pittsburgh. They had remeasured the proton's magnetic moment and obtained the result 2.47 ± 0.07. The two results, Rabi's and Stern's, were still at odds. For Rabi, there was no choice but to push on. "I want to make sure I've cleaned up in the phenomena I've seen," said Rabi. "I don't want anybody to make a great discovery in the field I'm working in. So, my idea was to do this as accurately as possible to see if we could get this fitted together and know the reasons behind it."[8]

The T-field introduced in the second series of experiments on the hydrogens was the sire of the magnetic resonance method. The T-field was a static field and it provided the means to establish the signs of the magnetic moments. Rabi used the T-field through 1936 and into 1937 for a number of experiments on other nuclei. There was no competition from other laboratories. Rabi's nonchalance was challenged in September 1937, when C. J. Gorter from the University of Groningen in Holland visited his laboratory at Columbia. Gorter asked Rabi why he wasn't using an oscillating magnetic field rather than the static T-field. Rabi had planned to do this eventually, but other work delayed the construction of new apparatus. "Well," said Rabi later, "I liked what we were doing, but I saw that he [Gorter] might go after it and we might get some competition. So I said, 'Let's do it.'"[9] Gorter visited on a Saturday. On Monday modifications to the apparatus were begun.

When a beam particle went through the static T-field, the particle experienced the effects of a rotating field whose rate of rotation depended on the speed of the beam particle. Fast particles saw a high rotation rate or high frequency; slow particles saw a slow rate or low frequency. Since beam particles have a range of speeds, each saw a different rotation rate, a different frequency. This made the T-field approach essentially qualitative in character. When the T-field was replaced by an oscillating field, the qualitative method became quantitative. The rate of oscillation was large enough so that every beam particle experienced essentially the same applied frequency. This allowed the experimenters to exercise control, a known control, over the beam particles.

Here is how it worked. Beam particles left the source and after passing through collimating slits, the beam particles passed sequentially through three regions. Region 1 was the first deflecting magnet. Region 2 contained a uniform magnetic field with the oscillating field superimposed on it. Region 3 was the second deflecting magnet oriented so that deflections were opposite to those deflections in region 1. From region 3, the beam particles entered the detector. Beam particles left the first deflecting magnet in a particular quantum state. If no quantum transition was induced by the oscillating field in region 2, the beam particles were refocused into the detector by the second deflecting magnet. If, however, the frequency of the oscillating field in region 2 stimulated beam particles to make a quantum transition, the beam particles were not refocused by the second deflecting magnet and thus did not make it into the detector and the detector signal decreased. The magnetic moments could be determined directly from a knowledge of the strength of the uniform magnetic field and the frequency of the oscillating field in region 2 required to decrease the detector signal. Both of these parameters could be determined with high accuracy and precision.

With the magnetic resonance apparatus functioning according to expectations, the third series of experiments began on the hydrogens in mid-1938 and they were collecting data by July. Some unexpected results came out of this series of experiments (see Chapter 13). To get maximum exposure for their work, Rabi and his associates gave preliminary results on the proton at a January 1939 meeting in Washington, D.C., deuteron results were announced at a meeting in Seattle in June, and the full-length paper containing the final results was published in *Physical Review* in September. In this paper, the proton magnetic moment was given as 2.785 ± 0.02 nuclear magnetons and the deuteron magnetic moment as 0.855 ± 0.006 nuclear magnetons. The uncertainty in these results was 0.7 percent (see Figure 12.2).

The experiments on the hydrogens over the period 1934–1939 contain all the ingredients of first-rate science. Often what distinguishes great scientists is their ability to pick significant problems to study. Although Rabi rejected the religious practices of his youth, his metaphor for doing science drew on his religious upbringing. Doing good physics was "walking the path of God." A challenging physics experiment was "wrestling with the Champ." A worthy physics problem "brought you near to God." The pro-

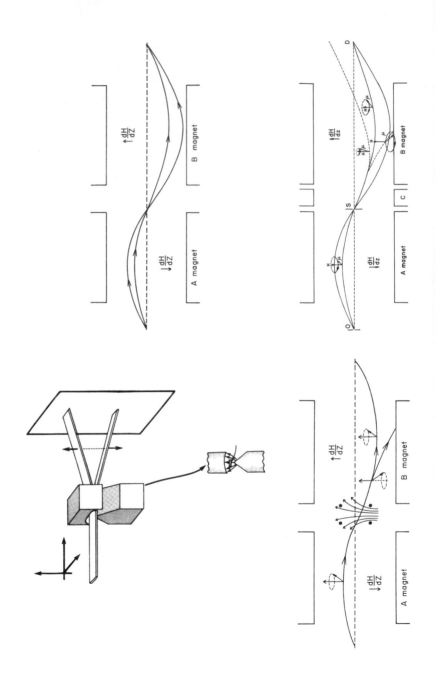

ton and deuteron, key players in the atomic world, brought Rabi near to God. "There is only one proton," said Rabi.

If the choice of a problem to study is sound, then the push toward ever-increasing accuracy, the quest for ever-increasing precision in the experimental results, is a hallmark of great science. From the first to the third experiment, the growing precision of the experimental results was stunning:

first experiment, 1934: proton 10 percent and deuteron 26
   percent uncertainties
second experiment, 1936: proton 5 percent and deuteron 4
   percent uncertainties
third experiment, 1939: proton 0.7 percent and deuteron 0.7
   percent uncertainties.

Rabi's experimental results challenged physical theory for decades. Great science coupled with high precision often leads to surprises, and sometimes important discoveries (see Chapter 13). Rabi expressed this well: "You felt if you were measuring the properties of hydrogen, the most fundamental nucleus, you just measure it and do it as well as you can. It was bound to fit into some or other scheme. And, if it didn't, it was significant. Here you have a system that you could understand. There were no complications. Anything I couldn't understand was because there was something to be discovered."[10] Finally, great experiments like Rabi's are characterized by an experimental design that is, in itself, a thing of elegance and beauty.

Figure 12.2 The evolution of the magnetic resonance method. The basic Stern-Gerlach experiment of 1922 appears in the upper left (see also Figure 11.1). Rabi's refocusing method of the mid-1930s with the A and B deflecting magnets is shown in the upper right. In the center of the lower left illustration is the T-field added between the deflecting magnets. The lower right illustration shows the magnetic resonance apparatus with a C magnet added between the A and B magnets. An oscillating field existed within the field of the C magnet.

The experimental methods developed over the period 1934–1939 ended with the magnetic resonance method. It was a powerful method that held within it the potential for physical and humanitarian applications. Shortly after World War II, the resonance method was extended in a form that is used today in physics, chemistry, and biology. Still later, the magnetic resonance method was applied to living organisms. Magnetic resonance imaging (MRI) is a noninvasive diagnostic tool that has become indispensable in contemporary medicine. And it all started back in 1934 in Rabi's laboratory.

As stated earlier, it is the nature of great research that one cannot predict what basic new insights about the workings of nature will emerge, nor can one dream of what practical applications will follow. The hydrogen atom has been particularly prominent in leading to new basic and practical outcomes. In a poignant moment, Rabi bore witness to an application of his own work in a situation he would rather have avoided and in a context he could never have anticipated.

A few months before his death in 1988, Rabi was hospitalized for a short time. His physicians wanted to understand his physical state as completely as possible, so Rabi was taken to the hospital's MRI facility and he was slowly rolled into the cylindrical magnet that plays a role something like the magnets in Rabi's molecular beam apparatus. In this magnetic field, the hydrogen atoms in Rabi's body could be detected by the magnetic resonance method and a cross-sectional image of his body generated, thereby providing physicians with detailed information about his physical condition. Once inside the apparatus, Rabi saw a distorted image of his face reflected in the shiny metal cylinder surrounding him. Rabi looked at his image. "It was eerie," said Rabi. "I could see myself in this thing. I would never have dreamed that my work would come to this."[11] A few weeks later, Rabi died peacefully at home.

13

New Nuclear Forces Required: The Discovery of the Quadrupole Moment of the Deuteron

Norman F. Ramsey and I. I. Rabi, 1939

Hydrogen has secretive as well as exuberant properties.

—Daniel Kleppner

Forces make things happen. Forces speed things up and forces slow things down. There are attractive forces and repulsive forces. An attractive force pulls magnetic decals to the refrigerator door and pulled Comet Shoemaker-Levy into the churning surface of the planet Jupiter; a repulsive force pushes two strips of scotch tape apart after they are stripped from a table top. In the arsenal of physical concepts, force is one of the most important.

The natural world pulsates with a diverse range of activities. Galaxies teem with violent, energetic outpourings; stars are born, evolve through different stages as they burn their supply of hydrogen, and when that supply is exhausted, they die; planetary systems form along with parent stars and, as we know from the Sun's family of nine planets, these stellar satellites exist in greatly different forms. Organic matter exists on at least one planet in the universe and does so in a staggering range of sizes, shapes, and dispositions; living organisms interact with their environments, altering each other as time advances. Below the visible, atoms and molecules do their tangos bringing together the chemical elements in a myriad of combinations to form and shape every mate-

rial object; and still further below the powers of our visual capabilities are the fundamental particles that seethe inside matter and
throughout the vacuum of the universe. This profusion of activity,
from galaxies to gluons, is the consequence of four basic forces
that individually and in tandem govern all of nature's processes
and cause everything that happens to happen.

The most familiar member of the four basic forces is gravity,
which, though the most feeble, reigns supreme in the universe at
large. The *gravitational force* binds galaxies into groups, grips stars
together in galaxies, crunches large, dead stars into black holes,
holds planets in orbit around their parent star, and keeps the feet
of children and their puppies firmly on the ground. The *electromagnetic force* shapes both the living and nonliving worlds, attract-
ing and repelling the atomic-molecular units that determine the
architecture of material objects. The electromagnetic force endows a chair with the ability to support a seated person against the
pull of gravity. The *weak force* governs certain behaviors of the basic particles. Unstable atomic nuclei that decay by the emission of
an electron, called radioactive β-decay, are an example of a process governed by the weak force. Finally, the strongest basic force
of the four is the force that holds the atomic nucleus together and
is called, simply, the *strong force*. The strong force is 10^{40}, that is,
10,000 trillion trillion trillion times stronger than the gravitational force. The gravitational force is weak relative to the other
basic forces, but it dominates the cosmos because the gravitational
force is always attractive and because stellar and galactic masses
are so enormous.

Although it is accurate to describe forces in terms of pushes and
pulls, physicists want to understand forces in quantitative detail.
For example, the gravitational force is an attractive force; thus, it
never pushes, it only pulls. But more important, physicists want to
know exactly what determines the strength of the gravitational attraction. Isaac Newton determined this and his result is learned

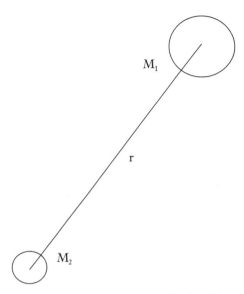

Figure 13.1 The attractive gravitational force pulls the two masses together along the line connecting their centers.

by every student of physics. The gravitational force between two objects increases in proportion to the product of their masses (M_1M_2) and decreases in proportion to the square of their separation ($1/r^2$). Newton went further. He demonstrated that the attractive force is a central force, which means it acts along the line connecting the centers of the masses. Knowing the masses of two objects and their separations, the magnitude of the gravitational force acting between them can be calculated exactly (see Figure 13.1). Because the gravitational force is known in quantitative detail, the appearance of comets and the coming of solar and lunar eclipses can be predicted with unerring precision. Physicists strive to understand each of the four basic forces in complete quantitative detail.

As soon as the atomic nucleus was recognized as a collection of protons (and eventually neutrons), physicists knew that a special new force was required. The reason is straightforward: each proton is positively charged and the electromagnetic force between the like-charged protons is repulsive. But nuclei exist in stable form, and since they do, a new attractive force is required to override the repulsive electromagnetic force and hold the nucleus together. The domain inside an atomic nucleus is an interplay between the electromagnetic and the strong forces.

Following James Chadwick's discovery of the neutron in 1932, and throughout the decade of the 1930s, the atomic nucleus became the frontier of physical research. And pushing the boundaries of this frontier were many American physicists.

By 1930, the young physicists who had gone to Europe during the 1920s to be where the new physics was being created had come home. During their sojourn in the European laboratories, physicists like Edward Condon, J. Robert Oppenheimer, and I. I. Rabi resolved among themselves to bring American physics out from under the dominating shadows of Niels Bohr, Werner Heisenberg, Erwin Schrödinger, Wolfgang Pauli, and the other leading European physicists. These young American physicists vowed to make American physics second to none. By the time the neutron was discovered in 1932, they were well on their way to fulfilling their vow.

Three things occurred in American laboratories in 1932: the positron was discovered by Carl Anderson at the California Institute of Technology, deuterium was discovered by Harold Urey at Columbia University, and Ernest Lawrence at the University of California, Berkeley extended the energy of the cyclotron to the 1 million volt level. Thus, the frontier of physics was shifting to American laboratories. Much of the nuclear research originated from those universities where the "boys," who had learned the new physics and had assimilated the spirit of world-class research in Europe, had come home to lead their own research groups.

Oppenheimer and Rabi are particularly noteworthy for their

impact on American physics. Both Oppenheimer and Rabi returned from Europe in the fall of 1929. Oppenheimer went to the west coast where, in a joint appointment between the University of California, Berkeley and the California Institute of Technology, he trained a generation of American theoretical physicists. Rabi established himself on the east coast where, at Columbia University, he started his molecular beam experimental research. By 1938, Rabi had an extraordinary group of young students and postdoctoral fellows who would eventually distinguish themselves in the world of physics. Among the group were four future Nobel laureates: Rabi himself won the prize in 1944; Polykarp Kusch won the prize in 1954; Julian Schwinger in 1964; and Norman Ramsey in 1985. In addition, Jerrold Zacharias would eventually hold the highest faculty position, university professor, at the Massachusetts Institute of Technology; Donald Hamilton became the head of the physics department at Princeton University; Sidney Millman rose to an executive position at AT&T's Bell Laboratories; and Jerome Kellogg became a division director at the Los Alamos National Laboratory.

Scientific research is an intensely personal, private, and competitive activity. Ideas take form in the quiet of an individual's mind. Honors come to individual scientists whose ideas determine the direction their science takes. At the same time, scientific research is strongly influenced by interactions among scientists and by group dynamics. As we shall see, both the personal and group aspects of research worked together in Rabi's group at Columbia.

When Rabi and his students started the magnetic resonance experiments on the hydrogens, it was assumed that the strong force acting between the protons and neutrons inside a nucleus was, like the gravitational force, a central force acting along the line connecting these nuclear building blocks. Such a force acting between particles in the nucleus would give the collection of protons and neutrons a spherical shape. That was the considered wisdom when the series of experiments on the hydrogens, begun in 1934, continued through 1938.

The objective of the 1938 experiments was to measure the magnetic moments of the hydrogen and deuterium nuclei as accurately as possible. In 1938, however, a new sense of promise inspired the members of Rabi's group as they prepared to apply the new resonance method to the hydrogens and to measure the magnetic moments of the proton and the deuteron to a new level of precision. Eventually, this objective was accomplished successfully, but not without surprises that led to new basic knowledge about the atomic nucleus.

When the resonance method was applied to measurements on the hydrogens—H_2, HD, and D_2—beginning in mid-1938, the expectation was that the detector would sweep out a single, relatively sharp resonance peak shaped like the curve in Figure 13.4. However, Ramsey, who was doing the experiment, found something very different for H_2 and D_2. For H_2 they saw a broad, jagged curve with no well-defined peak, whereas for D_2 they observed a single broad peak with weak, featureless wings (see Figure 13.3). At this time, Ramsey was seeking an experiment that would serve as his dissertation topic, for which he must be the sole author, as required by Columbia University. Since these strange features of the H_2 and D_2 resonances were thought to be effects of the apparatus, Ramsey was asked to study them.

Ramsey was a gifted student. He brought a keen sense of physical theory and a facility with mathematics to complement his talent as an experimentalist. The data in Figures 13.2 and 13.3 were taken by Ramsey at the beginning of his dissertation research. He soon found that with much-reduced oscillator power the H_2 curves became six separate, well-defined resonances—the first time a multilined structure had ever been observed in magnetic resonance. Encouraged by this unexpected result, Ramsey studied

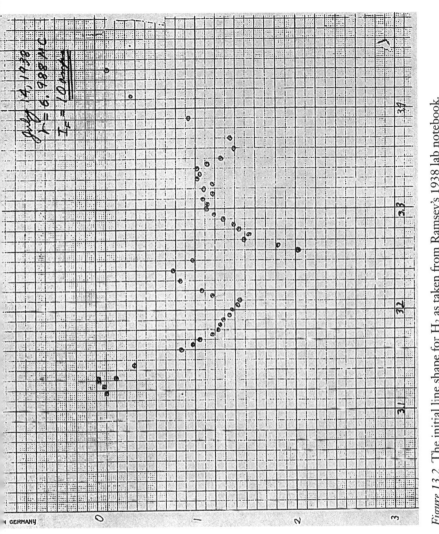

Figure 13.2 The initial line shape for H_2 as taken from Ramsey's 1938 lab notebook.

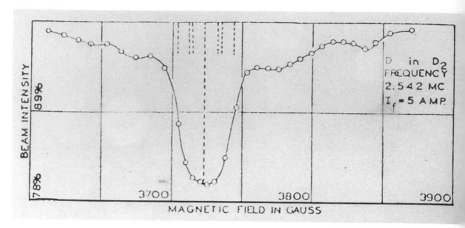

Figure 13.3 The initial line shape for D_2 as taken from Ramsey's 1938 lab notebook. These crude initial data should be compared with the refined later data shown in Figure 13.4 below.

the feeble wings of the D_2 resonance and discovered that these wings also revealed multiple resonances. Furthermore, these resonances were spread much further apart than those of H_2. This result was puzzling and indicated that there was something else going on in D_2. Ramsey was surprised.

Ramsey sent his data to Rabi, who was teaching at Stanford University during the summer of 1938, with a telegram: "What's doing this?" Ramsey kept exploring possible explanations. On September 14, he wrote in his notebook, "Can apparatus make the wings?" To resolve this question and to improve his data, Ramsey decided to redesign the apparatus. By this time, however, it was clear that something strange and possibly important was going on.

Rabi once said in reference to his research group, "We were an honest-to-goodness team." So it was that when Rabi returned from Stanford in mid-1938, the team nature of Rabi's research group asserted itself. Since Rabi sensed that something significant lurked in the peculiar signal shapes, he invited other members of his research team to participate in this potentially important experiment. As a result, Ramsey had to find another dissertation topic.

The next series of experiments began with a beam of deuterium molecules coursing through the newly designed apparatus. Methodical refinements were made from one experiment to the next, and in the process more and more specific details emerged from the "wings." By the end of the series of experiments, the details revealed were stunning. From the data shown in Figure 13.4 it is clear that the central large resonance is surrounded by satellite resonances—three on each side. It was the central resonance that gave the magnetic moment of the deuteron to high precision. The satellite resonance peaks required a new explanation.

To put the matter more accurately, it was not the satellite resonances themselves that were surprising, it was the large spread of these smaller resonances. The theory of the Rabi group predicted small resonances, but their predicted positions would have subsumed them under the large resonance and they would have been difficult, if not impossible, to observe experimentally. Something new was causing these satellite resonances to spread out and away from the central resonance. The spread was six times larger than their working theory predicted. The cause of this spread was a new and unexpected property of the atomic nucleus: the electric quadrupole moment.

The magnetic moment of the proton or deuteron makes it behave like a bar magnet with a north and south pole. This means that just as a bar magnet orients itself in a magnetic field, the proton or deuteron can align itself with a magnetic field. An electric quadrupole moment is a little harder to visualize. If the positive charge density within the deuteron concentrates slightly in two separate locations, then two other locations become, in relative terms, slightly negative, so there arise two centers of positive

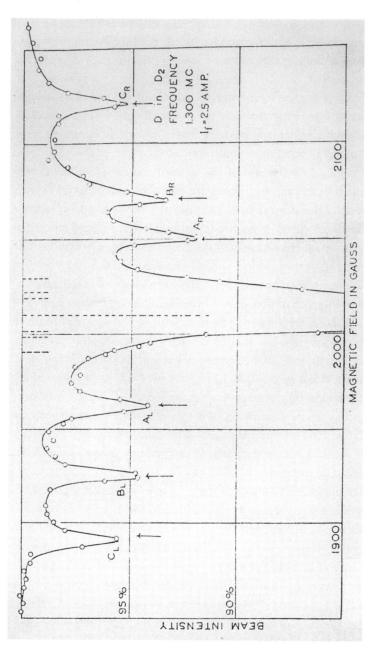

Figure 13.4 The resonance data for D_2 from which the quadrupole moment of the deuteron was determined.

charge density and two centers of negative charge density. Such a distribution of charge densities gives the deuteron an electric quadrupole moment. This is what Rabi and his students discovered.

The discovery of the deuteron's quadrupole moment in early 1939 was a complete surprise. "Indeed, when these experiments were started there was no question of investigating such a quadrupole moment, because current theory predicted, under the assumption of central forces between proton and neutron, that the deuteron . . . possesses . . . no electrical . . . quadrupole . . . moment."[1] With these words, the publication of Rabi and his group announced the discovery of the quadrupole moment of the deuteron.

This discovery required physicists to abandon the assumption of central forces acting within the atomic nucleus. Central forces were well understood and easily handled theoretically; thus, physicists were reluctant to abandon the idea of central forces. On November 28, 1939, J. H. Van Vleck from Harvard University wrote to Rabi, "I am not at all clear on just exactly the details of the set-up by which you are deducing this celebrated quadrupole moment. . . . If you can send me the details of just what you are doing, I can assure you I will do my best to punch holes in the attempted deduction of a quadrupole moment."[2] One week later, on December 5, Van Vleck wrote again to Rabi, "I have thought quite a bit about your experiments, but I cannot find any loop holes. . . . I congratulate you on the most interesting results of your experiments."[3]

The new force that replaced the central force was more complicated. A central force acting between two objects depends only on the magnitude of their separation. The presence of a quadrupole moment within the deuteron required a tensor force that depended not only on the separation between protons and neutrons, but also on the angles that their spins make with the line joining them.

In 1966, Hans Bethe of Cornell University, perhaps the most influential nuclear physicist of the twentieth century, identified three important events that occurred in nuclear physics between 1935 and 1948. First, he identified the detailed description of nuclear forces; second, the use of the Yukawa potential; and third, "and this is perhaps even more important—the quadrupole moment of the deuteron was discovered."[4]

Once again, hydrogen surprised physicists. A firmly held assumption about the nature of nuclear forces was squashed by nature's simplest atom and, in the process, it projected a new direction for physical theory.

14

Magnetic Resonance in Bulk Matter (NMR)

Edward M. Purcell and Felix Bloch, 1946

I remember, in the winter of our first experiments, just seven years ago, looking at snow . . . around my doorstep—great heaps of protons quietly precessing in the earth's magnetic field.

—Edward M. Purcell

The United States formally declared war on Italy and Germany on December 8, 1941. Long before the United States entered World War II, however, many American physicists were restive. Nuclear fission was discovered in Germany in late December 1938. Niels Bohr learned about this momentous event on January 3, three days before boarding the *Drottingholm* bound for the United States, where he was scheduled to visit the Institute for Advanced Study in Princeton for several months. Accompanying Bohr was his son Erik and a young colleague, Léon Rosenfeld, who spent much of the time during the trans-Atlantic trip listening to Bohr as he reflected and talked about the implications of the new discovery.

Bohr was met on the pier in New York City on January 16 by Enrico Fermi, his wife Laura, and Princeton physicist John Wheeler. Bohr spent the day in New York with the Fermis while Wheeler and Rosenfeld went on ahead to Princeton. Bohr had said nothing to Fermi about the fission discovery. He wanted Otto Robert Frisch and Lise Meitner, who proposed the fission explanation to explain Otto Hahn's experimental results, to get full credit before others heard it from him. Thus, on the train to Princeton, it was Rosenfeld who first shared the news of the fission discovery with Wheeler. Later that same day, the Princeton Physics Department Journal Club met and Rosenfeld gave a talk about the fission discovery. Ten days later, on January 26, a conference was held at George Washington University in Washington, D.C. where the fission news was announced to a broader audience. After the Washington conference the news about nuclear fission spread quickly. Almost as quickly physicists recognized that nuclear fission held the potential for a chain reaction with an attending release of energy—enormous energy.[1] This recognition was followed by a deep concern that this potential would be recognized and actualized by highly talented German physicists. Indeed, American physicists were restive.

The first real opportunity to respond to their unease had nothing to do with nuclear fission. The opportunity came by means of another trans-Atlantic ship, the *Duchess of Richmond*, which arrived in Nova Scotia on September 7, 1940. On this ship was a delegation of scientists from England. They carried with them what has been called the "most valuable cargo ever brought to our shores."[2] The cargo was small; it could be held in one hand. It was called a magnetron, invented by British scientists John Randall and Henry Boot, and it could produce 10,000 watts of power at a wavelength of ten centimeters.

On October 7, the magnetron was demonstrated for a group of physicists at the Bell Telephone Laboratories in New Jersey. The demonstration had a profound effect on those who saw it: this hockey puck–sized device was the means for developing small radar equipment that could be carried by fighter aircraft and ships.

Things moved rapidly after the magnetron showed its stuff at Bell Labs. Just a little over two weeks later, the decision was made to locate a federally supported laboratory at the Massachusetts Institute of Technology for the development of microwave radar.

The new lab was called the MIT Radiation Laboratory. This name was chosen to deceive: the University of California Radiation Laboratory was known throughout the world as a laboratory of nuclear physics, and it was hoped people would assume that the MIT Rad Lab was also devoted to nuclear studies, not radar. In 1940, despite the fission discovery, nuclear studies was still a subject only of academic interest. During the first week of December, physicists around the country hastily began to shut down their university laboratories. When their affairs were reasonably in order, they headed for Cambridge, Massachusetts, where over the next five years radar was developed to a high art. One of the first physicists to arrive in Cambridge was I. I. Rabi. Another physicist who arrived early was a young instructor from Harvard University, less than a mile up Massachusetts Avenue from MIT. That physicist was Edward M. Purcell. By late 1940, one year before the United States formerly entered World War II, American physicists had effectively gone to war.

In early 1943, American intelligence learned that the Germans were developing their own radar. In response, a new laboratory was established at Harvard to develop radar countermeasures to protect American war planes and ships from radar detection. Although this work began at MIT, the new center was moved to Harvard and called the Radio Research Laboratory. Joining the effort at Harvard in late 1943 was the Stanford University physicist Felix Bloch. Bloch had been at Los Alamos working on the bomb, but he did not like the military atmosphere at Los Alamos, where mail was routinely opened and surveillance was part of everyday life.

During the war, Purcell and Bloch were working in the same neighborhood. The countermeasures work at Harvard, however, was kept secret from the radar work at MIT. Thus, the two physicists saw each other only once during the war. Their one meeting occurred at a party celebrating Rabi's 1944 Nobel Prize.

Throughout the research and development efforts at MIT and

Harvard, new sources of radiation and new detection systems took their places in radar and counterradar equipment. In fact, a full array of new devices and new experimental techniques was brought to a highly refined state. New sources of microwave radiation were perfected; new electronic circuits were designed capable of detecting small signals. When the war ended, these radiation sources and detectors, these devices and techniques, these new electronic capabilities were suddenly available for research; thus, physicists stood on a threshold looking out on vast new fields of experimentation that could be conducted with the war-generated electronic circuits, the microwave sources of radiation, and the signal detectors. Physicists like Purcell and Bloch were quick to act.[3]

Purcell and Bloch were at different stages in their professional careers when they joined the war effort. Bloch, Purcell's senior by seven years, had been a professor at Stanford University since the spring of 1934 and had a significant body of research behind him when he started at Los Alamos. By contrast, Purcell finished his education just before he joined the MIT Radiation Laboratory and had just been appointed to an instructorship on the faculty at Harvard. Purcell had not established his own research interests.

The experience at the Radiation Lab was very important for Purcell's professional development. Thrown together with more senior physicists, he learned about their pre–Rad Lab work in physics. Particularly influential, as Purcell acknowledged, were the "physicists from Rabi's laboratory at Columbia."[4] Purcell learned from Rabi (to whom he reported), Ramsey, Zacharias, and Henry Torrey about Rabi's prewar magnetic resonance experiments, and as the war wore on, ideas began to take shape in Purcell's thinking.

Shortly before the war officially ended, a vast writing project was initiated by Rabi at MIT. The outcome of this effort was a series of books, twenty-seven volumes in all, that contained a complete account of the technical achievements of the Radiation Laboratory during the development of radar. These books, known as the Rad Lab Series, disseminated technical accomplishments to the larger scientific community and became textbooks for scientists and engineers during the years following the war. Purcell, with two co-authors, wrote volume 8 of the series.[5] The writing continued for several months after the war ended.

On a day in September, one month after V-J Day, fellow Rad-Lab physicists Robert V. Pound and Henry C. Torrey, also writing books about microwave electronics, asked Purcell to lunch with them. Torrey had been a student of Rabi's at Columbia before the war and was thoroughly conversant with Rabi's magnetic resonance method. As they were walking west along Massachusetts Avenue toward Central Square in Cambridge, Pound remembers that Purcell asked Torrey why the magnetic resonance experiment Rabi had carried out with molecular beams couldn't be done in solids. Purcell's idea was to take a solid rich in hydrogen, place it in a uniform magnetic field, which, according to quantum mechanics, would create quantized energy states, and then bathe these hydrogen atoms with radiofrequency radiation to see if energy is absorbed by the hydrogen atoms. Initially, Torrey expressed his skepticism about this possibility, but the next day, after an evening of calculations, he told Purcell that he was more optimistic about the possibility of a magnetic resonance experiment in bulk matter. Purcell invited both Pound and Torrey to join him in the effort.

About the same time, perhaps a little earlier, another walk took place. Bloch finished up at Harvard in late summer of 1945, but shortly before returning to Stanford, he went to lunch with Stanford colleague William W. Hansen, who was visiting MIT. Bloch and Hansen walked east along Massachusetts Avenue to a restaurant across the Charles River. During their lunch, Bloch told Hansen about an experiment he wanted to do when he returned to Stanford. Bloch's experiment was to take a bulk sample of a material rich in hydrogen, place it in a magnetic field, and see if the

Figure 14.1 Edward M. Purcell.

magnetic moments of the hydrogen nucleus could be reoriented by radiofrequency radiation. Though conceptualized and described differently, the experiment Bloch described to Hansen was equivalent to the experiment Purcell discussed with Pound and Torrey. Hansen was immediately interested in Bloch's idea and saw ways that the experimental approach described by Bloch could be improved.

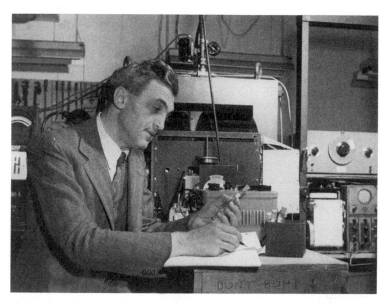

Figure 14.2 Felix Bloch.

Unbeknownst to each other, Purcell and Bloch started their individual experiments in the fall of 1945 at Harvard and Stanford respectively (see Figures 14.1 and 14.2). After the war there was little equipment and very little money at either Harvard or Stanford. Ernest Rutherford once said, "We have no money so we have to think." Both Purcell and Bloch followed Rutherford's admonition.

The Harvard experiment was a moonlighting effort, for Purcell, Pound, and Torrey were busy writing the MIT Rad Lab books. Both groups began scrounging for equipment. Purcell looked around the labs at MIT for a magnet, but found none. At Harvard, Professor J. Curry Street had once used a magnet in his cosmic ray work and Purcell arranged with Street to borrow that magnet, which was located in a shed behind the Lyman physics building. (The yoke of this particular magnet began its life as part of a generator of the Boston Elevated Railroad.) A signal generator was borrowed from Harvard's Psycho-Acoustic Laboratory. Bob Pound, an electronic virtuoso, took responsibility for the design of the electronic circuitry. The brass cavity that held the sample was made at MIT by a machinist whose work had slowed when the war ended. Torrey grappled with theoretical calculations to guide experimental design.

Meanwhile, Bloch had received $450 from Stanford for his experiment. Most of the money, $300, was spent on one instrument: an oscilloscope. They borrowed a magnet from the physics stockroom used for demonstration purposes in the classroom. Bill Hansen, also an electronics wizard, designed the electronics, and a graduate student, Martin Packard, put the circuits together.

The experiments of Purcell and Bloch bore the unmistakable influences of their past experiences. As a result, the two experiments were conceptualized in distinctly different ways. So different, in fact, that they initially had difficulty recognizing that the experiments were equivalent to each other.[6]

As stated earlier, Purcell was young and had no extensive body of research experience behind him. Nonetheless, Purcell's work had brought him face-to-face with the two-state quantum system. Perhaps the most telling instance of how the work of those around him influenced his experiments occurred during the radar work at the Radiation Laboratory. Rad Lab physicists were working on radar systems that operated at shorter and shorter wavelengths. When they reduced the wavelength to 1.25 centimeters, a curious malady plagued their results: sometimes the 1.25-centimeter systems worked and sometimes they didn't. Soon the cause of this fickleness was identified: the water content of the air. Harvard professor J. H. Van Vleck eventually showed that two quantum states of the water molecule had an energy separation just suited to absorb 1.25-centimeter radiation; thus, in moist air, the radar waves were absorbed and did not make it back to radar-system detectors.[7] The influence of this confrontation with the two-state quantum system can be seen in the experimental design Purcell adopted for his discovery of nuclear magnetic resonance (NMR) in bulk matter.

Bloch's experimental design was couched in more dynamical terms, which can be seen as deriving from his earlier work. Just before the war, Bloch and Luis Alvarez measured with great accuracy the magnetic moment of the neutron. In this work, the magnetic moments of neutrons were aligned by a magnetic field and then reoriented by an oscillating field. Bloch's neutron experiment bore similarities to Rabi's magnetic resonance method. Reorienting magnetic moments was the conceptual image that influenced Bloch's experimental design in his discovery of NMR.

Purcell's cavity was filled with two pounds of hydrogen-rich paraffin, the kind used to seal jars of jelly, which he purchased at a First National store on the way to his laboratory (see Figure 14.3). Bloch's sample cell contained a small amount of hydrogen-rich water. Both experiments focused on hydrogen: Purcell planned to look for energy absorption between two quantum states of the hydrogen atom; Bloch's intention was to look for a reorientation of the magnetic moments of the hydrogen nucleus.

Purcell's first experiment took place on Thursday evening, December 13, 1945. It was snowing hard. With their paraffin-filled cell mounted between the poles of their magnet, Purcell's team slowly varied the strength of the magnetic field looking for a response by their detector. From about 10:00 P.M. to 4:00 A.M. Friday morning, they kept at it: ranging back and forth across the magnetic field strength at which Torrey's calculations suggested energy absorption should occur. They found nothing.

The Purcell team reassembled early Saturday afternoon, December 15th. Purcell had arrived some seven hours earlier to turn the equipment on and to allow the paraffin sample to sit in the magnetic field for a time long enough (seven hours) to equilibrate the hydrogen atoms between the two quantum states. By late af-

Figure 14.3 The cell used by Purcell's group in their experiment. This cell is now at the Smithsonian Museum.

ternoon, they had seen nothing and were about to close down the apparatus when just "for kicks," as Pound recalls, they decided to turn the magnet up to full strength and watch the detector as they slowly decreased the magnetic field strength. As they watched the needle of the meter, they suddenly saw a sizable deflection. They went back and forth across this particular field strength and every time the detector responded. They had accomplished their goal: hydrogen atoms in paraffin were observed making a quantum transition from one energy state to another. Magnetic resonance in bulk matter was a reality.

Robert Pound kept very detailed records of the Purcell experiment. Such details are not available for the Bloch experiment. We do, however, know that their first attempt occurred during the fall of 1945 and that it was a failure. In a second attempt, shortly after

Christmas, they observed a blip on the oscilloscope as they decreased the strength of the magnetic field. This blip was evidence that Bloch and his colleagues were observing hydrogen nuclei in water molecules reorienting themselves in the external magnetic field. Magnetic resonance in bulk matter was a reality . . . on both the east coast and the west coast.

The Purcell group submitted their results for publication in a paper received by the editor of *Physical Review* on December 24, 1945.[8] The Bloch group sent their results for publication to the same journal and it was received on January 29, 1946.[9]

The discovery of magnetic resonance in bulk matter opened up a very active area of physical research. It has been applied to physical and chemical problems by physicists and chemists all over the world to the present day. In recognition of their work, Edward Purcell and Felix Bloch received the 1952 Nobel Prize in physics for their discovery of magnetic resonance in bulk matter.

In 1952, neither Purcell nor Bloch could have predicted the ways their discovery would advance understanding of solids, of the structure of chemical molecules, and even more. In fact, a representative from Dupont Chemical Company visited Purcell soon after the paper announcing the discovery was published. The Dupont scientist asked Purcell what the practical applications of NMR might be. Purcell responded that he could see no practical applications. In this, Purcell was very wrong (see Figure 14.4).

Nuclear magnetic resonance has become central in the work of chemists. The cover story of the November 5, 1984 *Chemical and Engineering News* dealt with NMR spectroscopy. In it NMR was portrayed as "a powerful and indispensable tool for obtaining new chemical information." NMR is a powerful tool for chemists because an atom like hydrogen, with a tiny magnetic moment in its nucleus, becomes a sensitive probe within a molecule. Norman Ramsey showed theoretically that the resonance frequency of a

Figure 14.4 Robert V. Pound, Henry C. Torrey, and Edward M. Purcell *(left to right)* at the retirement of Torrey from Rutgers University. This same photograph appeared on a poster celebrating the fiftieth anniversary at Harvard University of the discovery of NMR.

nuclear magnet is influenced by its electronic environment within the molecule. Specifically, the NMR signals originating from a hydrogen nucleus in one environment in the molecule are dramatically shifted from a hydrogen nucleus in another environment. From these "chemical shifts," detailed information about molecules can be determined.

The uses of NMR are manifold. NMR systems, for example, are lowered into bore holes by oil companies searching for oil deposits. More recently, NMR became a powerful tool in the hands of physicians. Unlike X rays and other methods used to see parts of the human body, NMR is completely noninvasive and benign and reveals vital information about bodily organs and their function. In the context of hospitals and medical facilities, the NMR method is called magnetic resonance imaging (MRI) and is a staple in medical practices.

NMR was a first-generation descendent of Rabi's discovery of magnetic resonance, whereas MRI is a second-generation descendent. As such, these practical consequences of Rabi's basic work demonstrate the unpredictability and value that are often inherent in pure research.

At a 1944 party in Cambridge where physicists from the Rad Lab and the Radio Research Lab gathered at Rabi's home to celebrate his Nobel Prize, Felix Bloch sat down at the piano and sang a little ditty:

> Twinkle twinkle Otto Stern
> How did Rabi so much learn?

When Bloch learned eight years later that he and Purcell had won the Nobel Prize for their independent discoveries of NMR, he once again displayed his wit and sent the following telegram to Purcell:

> I think it's swell that Ed Purcell
> Should share the shock with Felix Bloch.

15

Hydrogen's Challenge to Dirac Theory: Quantum Electrodynamics as the Prototype Physical Theory

Willis Lamb, 1947

The spectrum of the hydrogen atom has proved to be the Rosetta stone of modern physics.

—Theodor W. Hänsch, Arthur L. Schawlow, and George W. Series

Hundreds of physics conferences have been held since World War II, but with little doubt, the most influential conference of them all occurred on June 2, 3, and 4, 1947 on Shelter Island at the eastern end of Long Island. This conference occurred soon after physicists had returned from their wartime work designing the atomic bomb in Los Alamos, developing radar systems at MIT, or on other war-related projects in laboratories located around the country. Most of the physicists had returned to their prewar faculty positions and were looking ahead. It was an auspicious time for a conference.

The physicists invited to the Shelter Island conference met in New York City on Sunday, June 1 at the American Institute of Physics, boarded a battered old bus, rode across the length of Long Island, transferred to a ferry, and finally arrived at the Ram's Head Inn, where the conference was held.

The Shelter Island conference was designed to allow an elite group of physicists to assess their subject and to establish directions for further research. Only twenty-four mostly young physicists attended the conference. The structure of the meeting allowed ample time for free-flowing discussion. Three papers, designed to launch the discussions, were presented on the foundations of quantum mechanics. The significance and results of this conference exceeded all expectations. After the conference, J. Robert Oppenheimer wrote that it was "the most successful conference we had ever attended."[1] Years later, in 1966, Richard Feynman stated that "There have been many conferences in the world since, but I've never felt any to be as important as this."[2] And I. I. Rabi said that the conference will be remembered as the starting point of remarkable new developments.[3]

The meeting began on Monday morning with accounts of two recent experiments that reported small and subtle deviations in the energy states of the hydrogen atom. From that point on, the data from the hydrogen atom and their implications were a dominant topic of discussion. With the hydrogen atom, small things are significant. In part, this is because the hydrogen atom is so simple that even a small discrepancy has no place to hide. More to the point, the hydrogen atom is unique in that physicists can apply physical theory to it with exactitude: neither assumptions nor approximations compromise the outcomes of theoretical analyses of hydrogen. Therefore, with precise experimental data drawn from the hydrogen atom, physical theories must provide an accurate account of measured results or else the theory itself must be altered. This is what happened as a result of the Shelter Island conference: the results gave added impetus to the most successful theoretical framework physicists have produced, quantum electrodynamics (QED). The modern QED framework has shaped large bodies of physical research ever since the Shelter Island conference.

The experimental data presented at the Ram's Head Inn on

June 2, 1947 were tantalizing because of their accuracy and pro-
vocativeness. These data revealed small shifts in the energy states
of the hydrogen atom, thereby adding further to the observed
spectroscopic details of the hydrogen atom. The evolving spectrum of hydrogen demonstrates the way experiment and theory
goad each other and, in the process, provides a telling example of
how great science advances (see Figure 15.1 below).

At the turn of the century, the Balmer spectral series provided
physicists with hard data that cried out for explanation either in
terms of some established "gear works" or some less familiar dynamics working within the hydrogen atom. In addition to the four
Balmer lines—H_α, H_β, H_γ, and H_δ—was the discovery by A. A.
Michelson in 1891 that the spectral transition H_α was a doublet;
that is, it consisted of two closely spaced lines. These two H_α lines
were so close together that the Michelson interferometer, a high-
precision optical instrument, was required to observe it. In 1913,
Niels Bohr developed the first compelling model of the hydrogen atom. In this model, Bohr supplied the dynamics that were
indeed "less familiar." Bohr adopted a quantum condition when
he assumed that the hydrogen atom could exist only in certain energy states or levels. In other words, Bohr assumed that energy
was quantized. The Balmer series of spectral lines, as well as additional spectral series, resulted from transitions between these
fixed energy states. Bohr's model successfully accounted for the
gross features of hydrogen spectra. It failed to account for the
doublet nature of the H_α spectral transition.

Arnold Sommerfeld, in 1916, set out to explain the finer details
of the hydrogen spectrum. To do so, he refined Bohr's model by
adding the possibility of elliptical orbits to the circular electron
orbits Bohr had assumed in his model. This step alone, however,
was not enough as the energies associated with the elliptical orbits
were the same as those associated with the circular orbits. These
elliptical orbits, however, did take the orbiting electron on a wild

ride within the hydrogen atom as the electron swept closer to and roamed farther from the nuclear core than was the case with Bohr's circular orbits. It was when Sommerfeld brought relativity theory to bear on the issue that Bohr's energy states were split into additional energy states clustered closely together. This opened the way for additional spectral transitions that explained hydrogen's fine structure and brought experimental data and theory into harmony. However, in spite of this agreement, Bohr's model and Sommerfeld's extension of that model were *ad hoc* in character; that is, both were based on assumptions specifically made to account for hydrogen's spectral lines and thereby neither emerged out of a theoretical base of established general validity.

The creation of quantum mechanics in 1925 by Heisenberg and in 1926 by Schrödinger did provide a firm theoretical basis for the quantum nature of the hydrogen atom. Bohr's quantum condition was no longer *ad hoc:* quantization became a natural consequence of the wave-particle nature of the electron and all other subatomic particles. But neither Heisenberg's nor Schrödinger's quantum mechanics provided an adequate account of the details of the hydrogen spectrum.

To explain the spectrum of the hydrogen atom, all the facts had to be known and the power of both quantum mechanics and relativity theory were required. Neither version of quantum mechanics—Heisenberg's nor Schrödinger's—incorporated the theory of relativity. Neither Heisenberg's nor Schrödinger's quantum mechanics embraced the spin of the electron—a basic property of the electron discovered after Heisenberg and before Schrödinger did their seminal work.

In 1928, Paul Dirac brought quantum mechanics and relativity together and, in the process, provided a solid theoretical basis for understanding the spectrum of the hydrogen atom. He made no assumptions, working from the firm footings of quantum mechanics and relativity theory. His work not only gave a complete explanation of the hydrogen spectrum, but also accounted for the spin of the electron as well as the electron's magnetic moment.

The energy-level diagram for hydrogen, shown in Figure 15.1 complete with spectroscopic notation, has three columns that summarize the evolution of physicists' understanding of hydrogen's spectrum from Balmer and Bohr through Sommerfeld and Dirac. As shown, the energy states given by Dirac's 1928 theory are the same as those determined by Sommerfeld, but with the *ad hoc* character eliminated. This figure, not drawn to scale, shows the lowering of the energy states relative to Bohr's results as well as the splitting of the Bohr energy states due to both relativistic and magnetic effects. (The magnetic effects arise from the elec-

Figure 15.1 The energy states of hydrogen according to Bohr, then Sommerfeld and Dirac, then Lamb *(left to right)*. The left-hand column shows the energy states that resulted from Bohr's 1913 model of the hydrogen atom. Each state is labeled by the principal quantum number, n. The lowest energy state, $n = 1$, is called the ground state of the atom and is the state normally occupied by the atom. The ground state determines the physical size of the atom. States with higher energy are labeled by $n = 2$, $n = 3$, and so on. Transitions among these states are the origin of hydrogen's spectrum. The middle column shows the changes that result from relativistic and magnetic effects. The states are lowered in energy and additional energy states appear. Transitions among these states give rise to the "fine structure" of hydrogen's spectrum. Sommerfeld accounted for this complication, but his approach now has only historical interest. Dirac provided the correct explanation of these energy states by combining quantum mechanics and relativity theory. The right-hand column shows the departure from Dirac's theory that resulted from Lamb's experiment. The separation between the $2S_{1/2}$ and $2P_{1/2}$ states is called the Lamb shift. This energy-state splitting prompted refinements in quantum electrodynamics. Not drawn to scale.

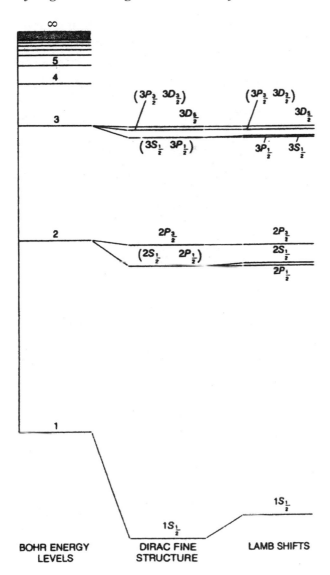

BOHR ENERGY
LEVELS

DIRAC FINE
STRUCTURE

LAMB SHIFTS

tron's spin and the attending magnetic moment, which interacts with the magnetic field generated by the relative motion between the charged electron and proton.) The five spectral transitions allowed between these five energy state levels are so close together in wavelength that in 1928 they could not be observed individually; rather, a doublet—two spectral lines—was observed for the H_α transition.

In 1938, new experimental data hinted that a small discrepancy might lurk in the spectrum of hydrogen. In Dirac theory, there are three energy states associated with $n = 2$, labeled $2S_{1/2}$, $2P_{1/2}$, and $2P_{3/2}$. But in Dirac theory, the $2S_{1/2}$ and $2P_{1/2}$ levels have the same energy. Two states with the same energy are called degenerate states. The new data, from experiments by W. V. Houston and R. C. Williams, hinted that these two degenerate states might, in fact, have slightly different energies.[4] Such a result would be at odds with Dirac theory. Other physicists obtained experimental results that appeared to be consistent with Dirac theory; that is, they observed each level as having the same energy. When World War II brought such experiments to a halt, the data were not conclusive. Dirac theory still seemed adequate and survived the war years intact.

The two reports that started the deliberations of the Shelter Island Conference in 1947 and dominated much of the ensuing discussions were given by Willis Lamb and Rabi. Both reports brought fresh data to the participants, data laid bare by the technology that came out of World War II. In the process, Lamb and Rabi exposed the possible discrepancy in Dirac theory that laid dormant throughout the world conflict.

In the twentieth century only a few practitioners excelled in both the theory and practice of physics. Willis Lamb was one of these. As an undergraduate chemistry major at Berkeley, Lamb came to recognize that it was theoretical physics that most interested him. He remained at Berkeley, but switched to physics upon entering graduate school. Eventually, he wrote a theoretical dissertation under the direction of J. Robert Oppenheimer. Soon after his formal education ended, the war came and Lamb, who had accepted a position at Columbia University, was recruited by the leaders of the radar project. Since the MIT Radiation Laboratory had a satellite laboratory at Columbia, Lamb stayed there. Lamb's Rad Lab assignment was to design, build, and test magnetrons with the objective of creating short-wavelength versions of the ten-centimeter device that had come from England and had started the radar initiative. In this capacity, Lamb honed his experimental skills. When the war ended, the microwave instruments created for war use awaited peacetime applications.

Lamb, of course, was thoroughly familiar with Rabi's molecular beam experiments, and Lamb's experimental design drew from the Rabi beam approach. The objective of Lamb's experiment is clearly expressed in a progress report he wrote in 1946:

> The hydrogen atom is the simplest one in existence, and the only one for which essentially exact theoretical calculations can be made on the basis of the fairly well confirmed Coulomb law of interaction and the Dirac equation for the electron. Such refinements as the motion of the proton and the magnetic interaction with the spin of the proton are taken into account in rather approximate fashion. Nevertheless, the experimental situation at present is such that the observed spectrum of the hydrogen atom does not provide a very critical test either of the theory or of the Coulomb law of interaction between point charges. A critical test would be obtained from a measurement of the fine structure of the $n = 2$ quantum state.

It was the $n = 2$ energy state that, in 1938, had hinted at a small energy difference between degenerate states. If it really existed, Dirac theory would be found wanting. Lamb set out to probe the

KEY ELEMENTS IN THE LAMB-SHIFT EXPERIMENT

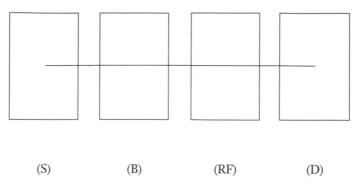

(S)   (B)   (RF)   (D)

Figure 15.2 The experiment that led to the discovery of the Lamb shift, as Lamb represented it in his Nobel Prize address. A beam of hydrogen atoms was directed from the source of the atoms (S) to a detector (D). In between, two regions were traversed by the hydrogen atoms, a bombarding region (B), and a magnetic field and RF region. Not drawn to scale.

two $n = 2$ energy states to determine whether their energies were identical or very slightly different.

There is an elegance and beauty to Lamb's experiment. In his experiment, shown in Figure 15.2 as Lamb represented it in his Nobel Prize address, a beam of hydrogen atoms was directed from the source of the atoms *(S)* to a detector *(D)*. In between, two regions were traversed by the hydrogen atoms. The first region was a bombarding region *(B)*, where electrons collided with the hydrogen atoms. The purpose of the bombardment was to excite a small fraction of the hydrogen atoms from their ground level, $n = 1$, to a higher energy level, $n = 2$, which was the state of interest. It so happens that one component of the $n = 2$ state, the $2S_{1/2}$ state, is a metastable state, which means that it has a relatively long life before decaying back to the $1S_{1/2}$ ground state. Lamb exploited this long-lived property, which allowed hydrogen atoms in this critical state to get through the apparatus and be detected before it decayed back to the ground state. After the hydrogen atoms were bombarded, they passed through a region where they were subjected to both a magnetic field and radiofrequency (RF) radiation, which, in combination, were designed to seek out the suspected energy discrepancy by stimulating transitions directly between the two close-lying states.

With hydrogen atoms in the desired $n = 2$ state moving through the apparatus, Lamb varied elements in the RF box to induce transitions directly between the metastable $2S_{1/2}$ state and the $2P_{1/2}$ state, the two states suspected of having different energies. If true, transitions would be induced between them and, once in the $2P_{1/2}$ state, the atom would immediately decay to the ground $1S_{1/2}$ state, with an attending decrease in the signal seen by the detector. That is exactly what happened. At a particular RF, the detector signal decreased. Knowing the RF, the energy difference between the $2S_{1/2}$ and $2P_{1/2}$ states was determined. The experimental result was unequivocal: there was an energy difference between two states that could not be accounted for by Dirac theory. The new energy states for $n = 2$, according to Lamb, are shown in the right-hand column in Figure 15.1. The separation between the two levels is called the Lamb shift.

What is the significance of the Lamb shift? The Balmer lines arise principally from the electrical interaction between the positively charged nucleus and the negatively charged electron. The fine structure as described by Dirac theory arises from relativistic effects. The Lamb shift is explained by QED. The explanation of the Lamb shift by means of QED is, as might be expected, rather subtle. The electromagnetic field must be expressed in quantum terms. Within the context of a quantized electromagnetic field, the complete explanation takes into account at least four factors. First, the electron interacts with its own electromagnetic field, which modifies the electron's magnetic moment (to be discussed in Chapter 16). Second, photons associated with the electromagnetic field can spontaneously produce pairs of electrons and positrons in the neighborhood of the nucleus. These negatively charged electrons together with the positively charged positrons effectively polarize the space surrounding the nucleus and shield the hydrogen's electron from its nucleus. Third, fluctuations in the vacuum surrounding the electron can alter its motion and its kinetic energy. Finally, there may be corrections involving the proton's size, charge distribution, mass, and motion. QED puts these elements together in a consistent way and provides a very accurate account of the Lamb shift. To appreciate the power of QED, consider the fact that the measured frequency of the spectral transition between the two levels resulting from the Lamb shift is 1,057.845 MHz and the value calculated from the basic equations of QED is 1,057.853 MHz. Experimental measurements and theoretical calculations agree to within ten parts per million.

Years later, when Willis Lamb celebrated his sixty-fifth birthday, Freeman Dyson wrote to him: "Your work on the hydrogen fine structure led directly to the wave of progress in quantum electrodynamics. . . . You were the first to see that that tiny shift, so elusive and hard to measure, would clarify in a fundamental way our thinking about particles and fields."[5] Dyson's remark, made in 1978, confirms Rabi's judgment of the Shelter Island conference thirty-one years earlier as "the starting point of remarkable new developments." One of these remarkable new developments was the full development of QED, which has become the paradigmatic theory of physics. Another one of the "remarkable new developments" was in response to the second report that started the Shelter Island conference: Rabi's report of new data on the hydrogen energy states. But that is a subject for the next chapter.

16

The Hydrogen Atom Portends an Anomaly with the Electron

I. I. Rabi, John E. Nafe, and Edward B. Nelson, 1946

As the simplest of the stable atoms, hydrogen permits unique confrontation between theory and experiment.

—Theodor W. Hänsch

The war changed I. I. Rabi's life and career. In his dual roles as the associate director of the MIT Radiation Laboratory and the director of research, Rabi worked closely with policy makers and military leaders. In the beginning, Rabi's discussions, particularly with military representatives, were infused with condescension, suspicion, and a benign refusal by generals and admirals to share vital military information. Rabi needed to know and understand the battle strategies being planned by the military in order to guide the development of radar equipment that would complement and support the war effort. But Rabi could play hardball himself, and soon his wisdom and intellectual toughness were recognized by Pentagon leaders. The unease dissipated, and frank, open discussions became the norm. This experience, coupled with his Nobel Prize in physics in 1944 and his close friendship with President Dwight D. Eisenhower, thrust Rabi into high levels of influence as the United States moved from World War II to the protracted Cold War.

Rabi's effectiveness during the war and the demands placed

upon him in the years that followed meant that his focus changed and his productivity in his own Columbia research laboratory never reached the same level of intensity as it had before the war. As Rabi said, "The prize opens doors for you, doors which perhaps should not be opened. It attracts you away from your field; it brings many distractions. I'm not a man to open doors and tell people, "Do this, or do that." If I am invited in, however, then I can go into action. The Nobel Prize caused me to be invited, it brought the invitation. If I hadn't won the Nobel Prize, I wouldn't have had the temptation to respond to them . . . nor would I have had the obligation."[1] Of course, Rabi could have made different choices, but he did not. Thus, Rabi's most significant postwar research results were those he reported in 1947 to the participants assembled at the Ram's Head Inn on Shelter Island.

The data Rabi presented on June 2, 1947 were based on results from a molecular beam experiment with hydrogen atoms. Before the war, Rabi's research on the hydrogens had focused on the measurement of the magnetic moments of the nuclei of the hydrogen atom: the proton and the deuteron. The experiment that Rabi and his two students, John Nafe and Edward Nelson, initiated when Rabi returned to Columbia University turned this prewar approach around. Because the magnetic moments of the proton and deuteron had been measured to high precision before the war by means of the magnetic resonance method, the new experiment took the values of the magnetic moments as a given and explored the energy states of the hydrogen atom themselves.

The two energy states that were the focus of Rabi's experiment were a direct consequence of magnetic influences at work within the hydrogen and deuterium atoms, arising from the magnetic moments of the electron, proton, and deuteron. The spectrum of hydrogen (and other atoms) originates from transitions between quantized *energy* states. However, in addition to energy, *angular momentum* is also quantized. Angular momentum arises from three sources: the orbital motion of the electron around the nucleus, the spin of the electron, and the spin of the nucleus. The quantized nature of angular momentum restricts the shapes of the orbitals allowed. The spins of both the electron and the nucleus give each an intrinsic magnetic moment. The electron and the nucleus can orient themselves only in certain ways relative to magnetic fields and can interact with each other. Each orientation corresponds to small differences in energy. The small splitting of the Bohr-type energy states because of relativistic effects and the electron's magnetic moment is called fine structure; the still smaller splitting arising from the influence of nuclear spins is called hyperfine structure. The complete energy level diagram of the energy states of hydrogen is shown in Figure 16.1.

The focus of the Rabi-Nafe-Nelson experiment was a transition between two hyperfine states in the ground state (the lowest $n = 1$ energy state) of the hydrogen atom; specifically the $F = 0$ and $F = 1$ states shown at the bottom of the right-hand column in Figure 16.1. Knowing the magnetic moments of both the proton and deuteron and assuming, on the basis of Dirac theory, that the magnetic moment of the electron was known exactly, the frequency separation of these two states could be calculated with precision. The calculated values were as follows:

frequency separation, hydrogen = 1416.90 ± 0.54 MHz

and

frequency separation, deuterium = 326.53 ± 0.16 MHz.

In their experiment, Rabi, Nafe, and Nelson used a beam of hydrogen atoms and measured the transition frequencies between the two hyperfine states. The measured values were as follows:

frequency separation, hydrogen = 1421.3 ± 0.2 MHz

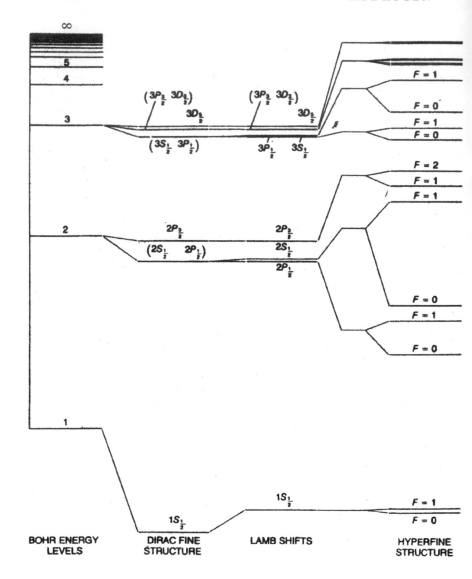

BOHR ENERGY
LEVELS

DIRAC FINE
STRUCTURE

LAMB SHIFTS

HYPERFINE
STRUCTURE

and

frequency separation, deuterium $= 327.37 \pm 0.03$ MHz.

In many experiments, these results would be regarded as confirmation of theoretical expectations. After all, the discrepancy between the measured and calculated values was about 0.26 percent. However, these results illustrate the power and beauty of hydrogen: nothing is hidden. If theory and experiment do not agree exactly, something is wrong. For more complicated atoms, larger deviations between theory and experiment must be tolerated. The hydrogen atom is uncompromising.

The paper that reported these results ended with the recognition that there was a problem: "Whether the failure of theory and experiment to agree is because of some unknown factor in the theory of the hydrogen atom or simply an error in the estimate of one of the natural constants, such as [the fine structure constant], only further experiment can decide."[2] This was the result that Rabi conveyed to the physicists at Shelter Island. Rabi's reputation as an experimentalist brought credibility to the measured results and issued a challenge to the theorists. As with the Lamb shift, it was quantum electrodynamics that was brought to bear on

Figure 16.1 The complete energy-level diagram of hydrogen showing fine and hyperfine structures. For an explanation of the first three columns, see the caption to Figure 15.1. The energy-state changes due to the interaction between the electron and nuclear magnetic moments are called hyperfine splittings. These splittings are smaller than the fine structure in the second column, but not as small as the splittings in Lamb shifts (third column). The two energy states in the lower right-hand corner are the two states on which the Rabi-Nafe-Nelson experiment focused and the source of the hydrogen 21-cm line. Not drawn to scale.

the small discrepancy and, in the process, QED exhibited its own potent power.

The culprit behind the disagreement between Rabi's experiment and Dirac's theory was the electron. Rabi's measurement was the separation of hyperfine energy states, which, as stated earlier, results from the interaction of the electron's magnetic moment with the proton's magnetic moment. Thus, the separation between these states was directly dependent on the magnitudes of both the electron's and proton's magnetic moments. Dirac theory had produced a value of the magnetic moment for the electron and its validity was, for all practical purposes, taken for granted. Such was the esteem for Dirac.[3] Yet, in spite of its incredible success, Dirac theory was incapable of fully describing the hydrogen atom. The theory neither explained the Lamb shift nor provided an accurate description of the electron's magnetic moment.

It was quickly recognized that the explanation for the disparity between data and theory resided in the assumed value, Dirac's value, for the magnetic moment of the electron. "It's something I should have seen right off," said Rabi, "but I didn't."[4] If the magnetic moment of the electron was slightly larger than that given by Dirac theory, harmony between experiment and theory could be restored. An experiment was designed by Polykarp Kusch, a colleague of Rabi's at Columbia, and Kusch's student Henry Foley to measure precisely the magnetic moment of the electron. The result was confirmed: the magnetic moment of the electron was slightly larger than predicted; hence, because it disagreed with Dirac theory, it was called anomalous. Given this experimental result, the challenge was to bring theory into agreement.

What is the electron? The electron, with both particle and wave properties, has four definite, quantitative properties: mass, charge, spin, and magnetic moment. Two of these properties, spin and the magnetic moment, seemed to be well accommodated by Dirac theory. (Why the electron has its particular charge and mass remains a mystery.) But there is more to the electron than meets the eye and it was quantum electrodynamics that revealed rather bizarre ways of conceptualizing the omnipresent electron.

So what *is* the electron? To answer this question, some insight into quantum electrodynamics is required. The principal actor in QED is the photon, which mediates the electromagnetic force. In the view of QED, the mechanism by which charges attract and repel each other is through the exchange of photons. As such, the electromagnetic field itself becomes quantized. The photon becomes the basic unit of the electromagnetic field. These photons have specific energies that are equal to $h\nu$ where h is Planck's constant and ν is the frequency of the photon. If atoms and photons exist together, they can interact with each other and atoms can absorb or emit photons.

A static charge, like an electron, takes on a new life in QED. An electron has an electromagnetic field consisting of quantized photons. Thus, the electron is surrounded by a cloud of photons. This cloud of photons surrounding an electron effectively reproduces the $1/r^2$ character of its measured electric field given by Coulomb's law. The electron can interact with its own electromagnetic field; that is, with photons in the cloud surrounding it. This interaction alters the behavior the electron would have in the absence of these interactions. To give a complete theoretical account of the electron interacting with its own field, corrections must be made by QED; in fact, by a new relativistic theory of QED. In the summer of 1947 Julian Schwinger recognized that to account for the results of the Rabi-Nafe-Nelson experiment, he would have to develop a relativistic QED, which he did with spectacular success during the next six months.

There is another consequence of the photon cloud around an electron. In this cloud of photons, the creation and annihilation of particles occur. It is these virtual particles, pairs of positive and negative particles, that lead to the polarization of the empty space surrounding the electron. Thus, the charge of the electron is partially screened from an outside viewer and, from a distance, appears slightly different from what it really is.

Incidentally, quantum electrodynamics transcends the electron. In other words, the ideas of QED go beyond the electron. For example, the concept of a basic interaction being mediated by an exchange of particles has been extended to both the weak and strong interactions with the mediating particles experimentally identified. The gravitational interaction is also assumed to be mediated by a particle, the graviton, but observing the graviton and quantizing the gravitational field has yet to be accomplished. The vacuum, again more generally, has taken on new life and it is viewed as teeming with particle creation and destruction.

The effect of the quantization of the electromagnetic field brings subtle changes to the energy states of the hydrogen atom as well as other atoms. These subtle changes are what Lamb measured. The QED corrections that were brought to bear on these changes accounted for the small shifts in the energy states.

So, given the character of QED, how did it bring an explanation to Rabi's data? Again, we ask: what *is* the electron? The electron does have a mass and a charge and, by virtue of the latter, the electron is the source of an electromagnetic field. The electron interacts with its own electromagnetic field and this interaction influences the mass and charge that is experimentally measured. The measured mass and charge are called the *physical* mass and *physical* charge. But what is the electron's mass and charge in the absence of these self-interactions? In the words of the trade, what are the *bare* mass and the *bare* charge? QED answered these questions and in the process revealed that the magnetic moment of the electron was slightly larger than that given by Dirac theory. It was this slightly larger magnetic moment of the electron that produced the unexpected deviation Rabi found in his data.

Stimulated by Rabi's data, experiments have been conducted to measure the magnetic moment of the electron with great precision. The precision achieved is astonishing. The electron's magnetic moment according to Dirac theory is

$$\mu = 1.0 \text{ Bohr magneton.}$$

The most precise measurement of the electron's magnetic moment is slightly larger:

$$\mu = 1.0011596521884(43) \text{ Bohr magnetons}$$

where the (43) means there is an uncertainty in the last two figures of ± 43. Richard Feynman put the incredible precision of this measurement into perspective when he wrote, "This accuracy is equivalent to measuring the distance from Los Angeles to New York, a distance of over 3,000 miles, to within the width of a human hair."[5]

The agreement that exists between the measured value of the electron's magnetic moment and the theoretical value calculated with quantum electrodynamics is noteworthy. QED, the most successful theory in physics, is the standard by which other physical theories are judged. The value of the electron's magnetic moment calculated from QED rather than Dirac theory is

$$\mu = 1.00115965214(4).$$

The agreement between the measured value of the electron's magnetic moment and the calculated value is 0.0001 part per million. As experimental methods become more and more refined and are capable of producing more and more accurate data, the world seems to get curiouser and curiouser. The natural world is much more imaginative than we ever imagined. When the imagination of humans held sway, human beings stood on a flat Earth positioned on the back of an elephant, which was supported on the back of a turtle, and so on. Instead, the Earth is a spinning ball hurtling around a nearby star and the human population stands on this globe—half of us upside-down relative to the other half. When human imagination was the guide, the matter of the universe consisted of earth, water, air, and fire. Instead, matter consists of atoms in motion, which in turn are made up of electrons, protons, and neutrons, and the proton and deuteron, in turn, consist of quarks and gluons. In addition, the so-called vacuum is pulsating with activity.

In arriving at insights into nature's bountiful imagination, we are fortunate that nature gave us the simple hydrogen atom. Its one electron with its nucleus of one proton or one deuteron has stimulated the feeble imaginations of scientists to probe behind the common-sense appearance of things and to soar to ever new heights of understanding. The concepts that have emerged from the laboratory have proven their power in synthesizing disparate realms of experience. At the same time, these concepts continue to challenge and boggle the best minds. As we look to the future, the hydrogen atom will continue to help us meet the challenge of embracing the natural world with understanding and, in the process, to understand better the place of humankind within the larger scheme of things.

17

Hydrogen Maps the Galaxy

Edward M. Purcell and Harold Ewen, 1951

> It was certainly apparent to [Jan] Oort right away . . . right away he saw it as a tool for learning something about the structure of the Galaxy.
>
> —Edward M. Purcell

Sometimes the most brilliant people can be short-sighted. Sometimes they can even be wrong. For example, consider two mathematicians who were contemporaries 200 years ago. One of them, the German Carl Friedrich Gauss (1777–1855), is universally regarded as one of the greatest mathematicians of all time. The other, Friedrich Wilhelm Bessel (1784–1846), also a German, was one of the best. Both men had similar opinions on the subject of astronomy and the importance of collecting certain types of data. Gauss said, "Speculation in astronomy first ceases and proper knowledge begins with the data which are capable of mathematical expression, such as the size and shape of the celestial bodies, their distances, their corresponding positions and most significantly . . . their motions."[1] Bessel noted, "Astronomy had no other task than to find rules for the motion of every star; its reason for being follows from this."[2]

Gauss and Bessel lived at a time when mathematicians doubled as physicists and both men knew that Newton had displayed the colors of the rainbow by passing the Sun's light through a prism. Yet neither saw the Sun's light or the light from other stars as a source of "proper knowledge" about celestial bodies. It's as if one sought to understand the workings of airplanes by determining their weight and shape, and by recording their position in the sky. In their pronouncements, Gauss and Bessel could not have been more wrong.

As Gauss and Bessel must have known, the Earth is awash with radiation from space. The most obvious, of course, is the visible light that radiates from Sun to Earth. Then there is the light from the twinkling stars that adorns the night sky. Invisible radiation was discovered between 1800 and 1801 in regions beyond both the red and violet ends of the Sun's visible spectrum. How far radiation extended into the long and short wavelengths beyond the visible regime was not known, but it certainly was established that the Sun's visible spectrum represented only a part of its radiation. Furthermore, it was observed in 1802 that the rainbow spectrum from the Sun had breaks, or dark lines, embedded in it which hinted that the radiation transmitted some type of information from Sun to Earth. So, a lot of things were happening around 1800—right under the intellectual noses of Gauss and Bessel.

By the middle of the nineteenth century, it was recognized that the radiation emitted by the Sun and other stars could provide information about their chemical composition through an analysis of their spectral lines. Our understanding of stars—how they are born, live, and die—has come from the classification of stars by virtue of the different types of spectra observed in the radiation that enters our laboratory apparatus from these far-off spangles in the night sky. From that time to the present, scientists have pushed toward investigating both longer and shorter wavelengths to determine what new information can be gleaned from stellar radiation (see Figure 17.1).

With good luck and serendipity acting in full force, a new window was opened to the universe in the early 1930s by a physicist at Bell Labs, Karl Jansky. Bell Labs was part of AT&T. AT&T had

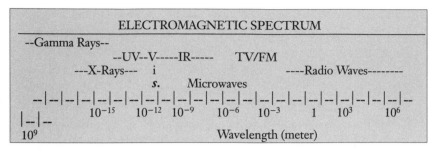

Figure 17.1 Sunlight is simply electromagnetic radiation that our eyes detect. The wavelengths in this fraction of the spectrum are shown as the vertical letters "Vis." This chart of the electromagnetic spectrum also shows waves with shorter wavelengths (ultraviolet, X-rays, and gamma rays) as well as longer wavelengths (infrared, microwaves, short radio waves (TV/FM), and longer radio waves).

adopted wavelengths at about fifteen meters in length for ship-to-shore and transatlantic communication. Unfortunately, there was interfering static that reduced the quality of communications, so Jansky was asked to examine this problem. Specifically, he set out to find the source of the static hiss—a challenging applied-physics assignment typical of those confronted by scientists in industrial laboratories.

Jansky discovered three sources of static: local thunderstorms (strong interference, but only during the storm), distant thunderstorms (weak static, but more constant), and finally a steady weak hiss of unknown origin. After a year of collecting data, Jansky concluded that the source of the weak static was extraterrestrial radio waves coming from the constellation Sagittarius, which almost coincides with the center of the Milky Way galaxy. Radio signals from space stimulated public excitement. After Jansky gave a talk on April 27, 1933 in Washington, D.C. entitled, "Electrical Disturbances of Extraterrestrial Origin," the *New York Times* ran a front-page story on May 5, 1933 under the headline, "New Radio

Waves Traced to Center of Milky Way." Americans tuned their
radios to a station that was connected directly to Jansky's antenna
and the radio announcer told listeners, "I want you to hear for
yourself this radio hiss from the depths of the universe."[3] Jansky's
1933 talk in Washington, D.C. is generally regarded as the beginning of radio astronomy.

As stated, good luck and serendipity were prominent in Jansky's
discovery. It just happens that the galactic center emits copious
amounts of radio waves at the particular range of wavelengths
Jansky was investigating. It just happens that the Earth's atmosphere is transparent to this same range of wavelengths so the signal reached his antenna. It also just happened that Jansky was
working during a sunspot minimum. At sunspot maximum, these
radio signals "from the depths of the universe" would have been
blocked by the Earth's ionosphere.

The response to Jansky's discovery of radio waves was muted in
the technical community because scientists regarded this new discovery as a curiosity. Few physicists followed up on Jansky's work
and astronomers showed little interest. Two reasons explain scientists' indifference. One reason was that the equipment needed to
examine radio waves accurately was lacking. This changed with
the end of World War II when advances in electronic devices, designed for the long wavelengths used in radar systems, became
available. The second reason was that Jansky's observation was
akin to what is heard when members of a symphony orchestra, all
sitting in their proper places, are warming up. A cacophony of
muted sound fills the concert hall as the instrumentalists independently go through their warm-up drills. Jansky's detected hiss was
the product of radio signals with a range of different wavelengths.
Like the orchestra warming up, there was no clean signal. This
too was soon to change.

As we learned in Chapter 1, most of the material in the universe
is hydrogen. Far and away, most of the interstellar gas is hydro-

gen. Those hydrogen atoms located near hot stars typically have their electrons stripped from them and they exist in ionized form. A small fraction of the hydrogen, about 1 percent, exists in molecular form—two hydrogen atoms chemically bonded together. The rest are individual hydrogen atoms, approximately one in every cubic centimeter throughout interstellar space.

Shortly before World War II ended, a Dutch astronomer, H. C. van de Hulst, predicted that these atoms of hydrogen might be detected by looking for a particular telltale signal. This signal would have a frequency of 1,420.4026 megacycles per second (Mc/s) or a wavelength of 21 centimeters (cm) and would arise from a particular spectral transition between two states of hydrogen.

This particular quantum transition later rose to stardom in the world of physics. It was the same transition that Rabi and his students, shortly after the war, found to be at odds with Dirac theory, and it was this transition that led to a new value for the electron's magnetic moment. By 1947, this transition had gained prominence from an experiment on earthbound hydrogen atoms; soon this same transition would assume galactic significance. Also, this same transition would become the basis for the most accurate atomic clock, which was developed in the late 1950s (discussed in Chapter 18).

A few years after van de Hulst's prediction, scientists in Holland began to prepare an experiment to determine whether, in fact, this spectral transition with its wavelength of 21 cm could be detected from hydrogen atoms located in the depths of space. Unfortunately, a fire in their apparatus set them back and, with the passage of time, their opportunity for the original discovery was lost. As we shall see, however, fate eventually smiled on the Dutch group.

In the meantime, Harvard physicist Edward M. Purcell and his graduate student Harold I. ("Doc") Ewen were setting up an ex-

Figure 17.2 The horn antenna used by Purcell and Ewen in their discovery of the hydrogen 21-cm line.

periment to scan the sky for this hydrogen signature—the particular wavelength of 21 cm. Neither Purcell nor Ewen knew that the Dutch group was in the process of preparing the same experiment.

The Harvard experiment was carried out with meager funds. The budget of $500 was spent to construct a plywood antenna that had a rectangular cross-section and expanded from its neck like the bell of a strange-looking musical instrument. The antenna was designed to receive and detect the 21-cm line from hydrogen atoms in the extreme cold of outer space. (The original telescope constructed at Harvard is shown in Figure 17.2. It is exhibited at the National Radio Astronomy Observatory in Green Bank, West Virginia.) The rest of the equipment was borrowed. By the spring of 1951, the experiment was ready to go.

Records do not exist that reveal how many fruitless nights were spent trying to detect the 21-cm line. In fact, at one point Purcell and Ewen decided to abandon hope of finding the line, but continued in order to make the negative results sufficiently convincing. It was during this period, on the night of March 25, 1951, that the hydrogen line was first detected. The signal came from hydrogen atoms in the constellation Ophiuchus located toward the center of the Milky Way. The source of this first signal was a cloud of hydrogen atoms extending across some 3,000 to 5,000 light years.

With the discovery of the 21-cm line of hydrogen, the young field of radio astronomy entered a new era. Suddenly, it was as if a maestro had arrived on the podium, the cacophony ceased, and a sweet tone, pure in character, came from the orchestra playing as one. In "Doc" Ewen's horn-shaped antenna the static hiss of random radio waves from outer space was replaced by a pure tone from hydrogen atoms. The pure tone had a frequency "around" 1,420 Mc/s with a wavelength of 21 cm.

It just happened that van de Hulst was at Harvard as a guest lecturer at the very time the experiment succeeded. Also at Harvard, by happenstance, was Frank J. Kerr from the radio astronomy group in Australia. Kerr tells what happened: "Purcell called us [van de Hulst and Kerr] all together on the morning after the overnight discovery. Cautious scientist that he was, he wanted to see confirmation of the detection before publishing the result. He suggested that van de Hulst and I should cable our respective institutions in Holland and Australia to report the discovery and ask whether early confirmation would be possible."[4] Purcell and Ewen sent an announcement of their discovery to the journal *Nature*, but Purcell asked the editors to hold up publishing the result until the Dutch and Australian groups could reproduce the results. The Dutch group was able to confirm the results quickly. The Australian group verified the finding in six weeks.

The results of the discovery were eventually published from the three groups—American, Dutch, and Australian—simultaneously in back-to-back papers in the September 1, 1951 issue of *Nature*.

Such was the character of Edward M. Purcell. How many scientists, one might ask, would delay publication of results so that others could share in the glory? Purcell was an unusual physicist, displaying none of the ruthlessness often found in men and women of ambition. Instead, kindness and consideration for others emanated from him. "Doc" Ewen also had to have been part of the decision to share the honor with the Dutch and the Australians. Ewen was a little more senior than the typical graduate student since he had served in the Navy during World War II. In the Navy, "Doc" was an instructor of celestial navigation. One of his students was Ted Williams, the great Boston Red Sox baseball player, who was a Navy pilot. During the Purcell-Ewen experiment, Ted Williams visited the Harvard physics laboratory to see "Doc" and to see the experiment firsthand. As Purcell later reported, the famous physicists at Harvard "were all aflutter" with the great Ted Williams in their midst. Whether it was his maturity, his service in the Navy, or the influence of Purcell, Ewen also deserves credit for sharing the honor of the discovery.

Both Purcell and Ewen knew their discovery was very important. One indication of its importance is that Purcell and Ewen have been incorrectly credited with winning the Nobel Prize for this work. (Purcell won the prize for his discovery of NMR.) No one can say, but it is possible their unselfish act of sharing the discovery with the other two laboratories deprived them of added honor. It is interesting to note that a year before he died, Purcell confided to one of his most illustrious students, Nicolaas Bloembergen, that he regarded his contributions to radio astronomy to be at least as significant as his discovery of nuclear magnetic resonance.[5] The discovery of the hydrogen 21-cm line opened possibilities for gaining new knowledge about our galactic neighborhood as well as the universe far from us—detailed and surprising knowledge.

The Sun with its nine planets is about 30,000 light years from the center of a large complex of stars we call the Milky Way—our galaxy. In spite of being a resident of this galaxy, we knew little about its structure, for it was rather like trying to determine the shape of a forest while standing among the trees. In addition, interstellar clouds of dust and hydrogen block visible light so scientists were unable to observe structural characteristics of the central regions of the Milky Way with optical telescopes. Radio waves, by contrast, are not blocked by these clouds. In fact, these interstellar clouds of hydrogen can be mapped by the 21-cm line.

The spiral arms of the Milky Way are composed mostly of hydrogen. Except for the stars within these spiral arms, these spiral arms themselves are invisible to optical telescopes. The 21-cm line of hydrogen makes structural details of these spiral arms "visible." Radio antenna tuned to this line can be swept across the disk of the Milky Way so that its spiral-armed structure can be observed. One such image constructed from hydrogen data was created by Gart Westerhout. It shows structural details of our galaxy (see Figure 17.3).[6]

How did the data permit Westerhout to do this? The nineteenth century mathematicians Gauss and Bessel thought the only proper way to study celestial bodies was to establish their motion. Gauss and Bessel were partly right. The hydrogen 21-cm line allows motions—radial motions—of hydrogen clouds, moving either toward or away from the earthbound detector, to be determined in a very precise way. Earlier the frequency of the observed hydrogen 21-cm line was given as "around" 1,420 Mc/s. This is because the observed frequencies actually clustered around the central frequency of 1,420 Mc/s. The "clustering" is a result of the motion of the hydrogen atoms. In the same way light from distant galaxies is red shifted from the expansion of the universe,

Figure 17.3 The Milky Way as seen by the hydrogen 21-cm line.

so those hydrogen atoms moving away from the radio antenna are shifted to slightly longer wavelengths (red shifted) and those moving toward the antenna are shifted to slightly shorter wavelengths (blue shifted). It is from these Doppler-shifted data that both the motion and the location of the Milky Way's spiral arms can be determined rather precisely.

If a magnetic field existed in interstellar space along the spiral arms of the Milky Way, the 21-cm line of hydrogen would be split into spectral components. This splitting is called the Zeeman effect. Thus, the 21-cm line of hydrogen held the potential of providing further information about the Milky Way. In 1968, Gerrit Verschuur discovered the presence of interstellar magnetic fields by detecting this splitting. In fact, it was found that the hydrogen clouds are permeated with magnetic fields.[7]

Beyond the Milky Way, the Clouds of Magellan, the galaxy closest to our own, have been observed by means of the hydrogen line. To the surprise of astronomers, this galaxy is bigger than it appears when viewed through optical telescopes. The red and blue shifts observed with the hydrogen 21-cm line further reveal that the Clouds of Magellan are rotating.

The discovery of the 21-cm line from hydrogen was the beginning of spectroscopy at radio frequencies in outer space. Following the Purcell-Ewen discovery many molecules have been found to exist in space. The water molecule, H_2O, and ammonia, NH_3, were discovered in 1969 by scientists at the University of California, Berkeley. The first organic molecule, formaldehyde, H_2CO, was also discovered in 1969. It was unexpected that three- and four-atom molecules such as these could exist in the barrenness of interstellar space. By 1970, it became clear that a whole domain of chemistry was active in the dark reaches of space. Perhaps the most unexpected molecular discovery was made in 1994 when the simplest amino acid, glycine, NH_2CH_2COOH, was discovered in interstellar space. Amino acids are the building blocks of life and to discover that these complex molecules exist in the cold vacuum of space has been one of the biggest surprises in the profound legacy of the 21-cm line.[8]

The hydrogen atom burst open the field of radio astronomy. Today, radio astronomical observatories are located at strategic sites in many nations of the world in both hemispheres. Scientists in these observatories are actively listening to the radio signals emanating from sources within the Milky Way and from the further depths of space. Many sources of radio waves have been identified that build our knowledge of the greater universe.

Hydrogen atoms are ubiquitous throughout the physical universe. Where these atoms exist in neutral form, such as between stars and perhaps between galaxies, they have the potential of emitting a 21-cm radio wave. On average, a hydrogen atom emits a 21-cm photon every 133 years. If we assume that there is approximately one hydrogen atom in each cubic centimeter of space, this means that the characteristic signature of the hydrogen atom leaves each and every cubic meter of space approximately sixty times each and every year. Some of these radio photons are emitted in the direction of the Earth, and when scientists choose to listen, their instruments hear a special form of music revealing information that only nature's provocative hydrogen atom can provide.

18

The Hydrogen Maser: A High-Precision Clock

Norman F. Ramsey and Daniel Kleppner, 1960

This is about the stillness in moving things,
In running water, also in the sleep
Of winter seeds, where time to come has tensed
Itself, enciphering a script so fine
Only the hourglass can magnify it, only
The years unfold its sentence to the root.

—Howard Nemerov, *Runes*

Time mystifies—it goads the imagination, and always has. The earliest records of human existence document that even our distant ancestors were aware of the passage of time. Ancient frescos on the walls of caves in France, dating from approximately 32,000 years ago, represent an act in the present depicting scenes from experiences in the past that scholars think were attempts to influence events in the future. It is reasonable to assume that the imagination that spawned these paintings was at work in the minds of early people long before their thoughts found expression on cave walls. Indeed, a recognition of time—past, present, and future—appeared very early in the human experience. These early humans also had a firm grasp of the temporal nature of life and they anticipated a life after death. Gravesites reveal human remains accompanied by artifacts designed to serve the deceased in the arena of the spirit.

Early in a child's life, he or she discovers the facts of birth and death. As children mature, their conscious awareness of mortality becomes more pronounced. Perhaps it is the facts of birth and death that invest the human mind with a preoccupation with time. Time carries with it the reality of human mortality and in response to this grim inevitability, religions provide their followers a route to serenity beyond the grave. The typical human's aspiration is captured by the cartoon character who said, "Listen, I don't want to live forever, but I damn well don't want to be dead forever either."

In addition to the metaphysical and religious implications that thinkers through the ages have brought to the concept of time, their awareness of time had very practical outcomes. Calendars, derived from observations of celestial bodies, were developed to track the long-term passage of time. A means to account for time in the short term was also an early need. The first clock was undoubtedly the Sun. As early as 3500 BC, Sumarians introduced manmade temporal divisions of time: the day and the year. Still using the Sun, Egyptians were measuring time with shadow clocks in 2000 BC. By 1600 BC, Egyptians had developed water clocks.

Robert the Englishman wrote in 1271 about an attempt to construct a weight-driven clock: "Clockmakers are trying to make a wheel, which will make one complete revolution for every one of the equinoctial circle, but they cannot quite complete their work. But, if they could, it would be a really accurate clock and worth more than an astrolabe or other astronomical instrument for reckoning hours."[1] When Galileo discovered that the period of a pendulum is independent of its amplitude, the pendulum became the basis for clocks and the seventeenth century witnessed great advances in timekeeping. As the accuracy of pendulum clocks improved, small effects from temperature and pressure became apparent and efforts to account for these small effects began the quest for enhanced accuracy.

The age of global exploration as well as global trade provided a very cogent motivation for designing and building clocks that were both rugged and accurate. The great explorers Vasco da Gama, Vasco Núñez de Balboa, Ferdinand Magellan, and Sir Francis Drake sailed blind, so to speak, as they had no way of knowing where they were located on an east-west line. In 1714, the English government offered a prize of £20,000 for a clock that would keep time with an error of not more than two minutes in forty-two days. The goal was to know one's location at sea, which required a knowledge of both latitude and longitude. Latitude could be determined from the Sun and stars, but longitude required an accurate knowledge of two times: the time at home port, for example, and the local time at sea. The compulsion to build accurate timekeeping devices, then and now, has been driven by very practical needs.

Time is one of the most basic concepts of physics. Motion is a play on the stage of space and time. To understand motion and to know where a moving object was in the past as well as where it will be in the future requires data that bring together specific spatial positions at particular times. For Isaac Newton, time had an absolute character: time had a real, universal existence unto itself and flowed inexorably forward from past to future, independent of events and influences that occurred *in* time. Newton's concept of time was consistent with human experience. However, this view changed dramatically with Einstein's theories of special and general relativity. The special theory of relativity effectively denied time its absolute nature because the theory showed a linkage between time and space. For example, clocks set up at quarter-mile intervals along a highway will, when synchronized, all read the same time to a physicist standing on the highway; however, to a physicist moving very rapidly along the highway, each clock reads a different time. Or, a clock observed to keep time accurately by someone sitting beside it is seen to run slowly by someone moving relative to the clock. Effects such as these have been confirmed in numerous physical experiments.

Einstein's general theory also brings new wrinkles to the idea of time. The generalized form of relativity predicted that clocks also are influenced by gravitation: the stronger the gravitational potential, the slower a clock will run. This prediction was confirmed in a beautiful experiment by Robert V. Pound and Glen Rebka in 1960, when they demonstrated the lengthening of the wavelength of a gamma ray traveling between the basement and the top of the physics building at Harvard University. This lengthening wavelength is equivalent to a slowing clock.

Einstein's theories of relativity were created during the early years of the twentieth century. The desire to confirm the many predictions of relativity has been one motivation in the search for ever more accurate clocks. The hydrogen maser clock is a part of that tradition.

The first clock based on atomic properties was driven by a crystal of quartz. Strictly speaking, a quartz clock is not dependent on the properties of the individual atoms of silicon or oxygen that make up the quartz; rather, this clock relies on the mechanical properties of a single crystal of quartz. In the late 1920s, scientists at Bell Laboratories designed a clock based on the quartz-crystal oscillator. Just as Galileo's pendulum oscillates back and forth with a fixed frequency and, in so doing, chronicles the passage of time in definite increments equal to the period of the pendulum swing, so the regular vibrations of a quartz crystal, occurring at a fixed frequency, can be applied to the keeping of time. In the 1940s quartz clocks replaced pendulum clocks as the standard of time.

Quartz clocks are accurate to about 0.0001 seconds per day. Such accuracy is more than adequate for everyday affairs. Suppose, however, the challenge is to measure the difference in clock rate between sea level and the top of Mount Everest? The answer is that a clock on the top of Everest loses about 0.000030 second over the course of a year relative to a sea-level clock. Quartz clocks could never capture this elusive time difference. Questions arising in physics required better timekeepers to learn their answers.

In 1945, I. I. Rabi was the Richtmeyer Lecturer at the annual meeting of the American Association of Physics Teachers that took place in New York City. In his lecture, Rabi announced the possibility of atomic clocks. This was picked up by the *New York Times* science reporter William L. Laurence. "'Cosmic Pendulum' for Clock Planned," read the headline of the article that appeared in the Sunday, January 21, 1945 edition of the paper. There followed three other headlines: "Radio Frequencies in the Hearts of Atoms Would Be Used in Most Accurate Clock" was the first. World War II was not yet over and the first paragraph of Rabi's prepared talk reveals the state of mind of physicists in early 1945: "I am very sorry that I will have very little that is new to contribute which has not already appeared in *Physical Review* or the *American Journal of Physics*. The war has taken care of that, but it may not be remiss to review a subject if only as a sop to our nostalgia for our peacetime physics."[2] From this beginning, Rabi went on to describe the magnetic resonance method that he and his students had developed prior to the outbreak of the war. He recognized the potential, inherent in this work, for a state-of-the-art timekeeper and he and his students talked about such a possibility before the beginning of World War II. Norman Ramsey remembers discussing an experiment with one molecular beam apparatus on a mountaintop and another in a mine to measure the gravitational effect on time. Ramsey, a graduate student at the time, said he recognized that he would get assigned to the gloomy mine while Rabi and others more senior would get the mountaintop view. Rabi's 1945 talk in which he proposed a new basis for an accurate clock—an atomic clock—was a continuation of discussions started over five years earlier.

Atomic clocks have their roots in Rabi's magnetic resonance

method. The reasons for this begin with the characteristics of an ideal clock. A clock must have an inherent period, for example, the time for a pendulum to swing through one cycle. An ideal clock is one whose intrinsic period is reproducible anywhere, any-time. A pendulum clock is not ideal because the period of the pendulum bob is affected by the clock's motion as well as by its altitude.

When an atom makes a transition from a high-energy quantum state to a lower energy state, electromagnetic radiation with a definite frequency and a definite period is emitted. When properly detected, this frequency, or period, becomes the ticking of an atomic clock, just as the crystal vibration frequency and the swinging frequency are the inaudible ticks of a quartz clock and a pendulum clock. The frequency emanating from the atom, however, is much less influenced by environmental factors such as temperature, pressure, humidity, and acceleration than are the frequencies from quartz crystals or pendula. Thus, atomic clocks hold inherently the potential for reproducibility, stability, and accuracy.

To utilize the radiation emitted by an atom for timekeeping purposes, its frequency oscillations must be counted, which is equivalent to counting the "ticks" of the atomic clock. In the magnetic resonance method, quantum transitions occur with frequencies in the microwave region, which could be counted with the technology available in the late 1940s. By contrast, the frequencies associated with the Balmer series of the hydrogen spectrum are in the optical region and have much higher frequencies. The technology to count the high-frequency oscillations associated with optical transitions was crude to nonexistent. Thus the magnetic resonance method was an ideal basis for atomic clocks.

In 1950, Norman Ramsey, a member of the team that developed the magnetic resonance method, made a basic modification to Rabi's molecular beam apparatus that significantly enhanced the accuracy of the molecular beam magnetic resonance method, but also portended enhanced accuracy for atomic clocks. Ramsey's modification was simple, but profound. In the original Rabi apparatus each beam particle passed through one oscillatory field. The accuracy of the measured results depended on how long particles remained in the oscillatory field. When Ramsey was setting up his laboratory at Harvard after World War II, he tried to lengthen the oscillatory region so that moving beam particles would spend more time in the field and thus increase the accuracy of his experimental results. This approach introduced insurmountable problems. In response, Ramsey had a brilliant idea: have each beam particle pass through two oscillatory fields separated along the beam path. This approach effectively kept beam particles "in the oscillatory field" for a longer time and worked beautifully. The Ramsey separated oscillatory field method has been used in many applications, including atomic clocks.

The first atomic clock based on the magnetic resonance method was built by Jerrold Zacharias, a longtime associate of Rabi's and a member of the MIT faculty after World War II. In 1952, Zacharias was teaching a seminar he had created for bright MIT seniors. During one class, they were discussing relativity theory and how some of the predictions of the theory could be tested with an accurate clock. Zacharias had been part of the prewar discussions in Rabi's laboratory about atomic clocks and was fascinated by molecular beams. The students had a clever idea for keeping beam particles in the oscillatory field for a longer time: design a vertical beam system and shoot particles up against gravity. After some time, the particles would fall back into detectors. Zacharias became fascinated with the idea of an atomic fountain and developed plans for two clocks using cesium as the beam atom: a small clock and a large fountain clock. In Zacharias's time the fountain clock never materialized, but the small cesium clock not only worked, it became a standard.

Zacharias gave his first report on his cesium clock in 1955.[3] In this clock, a beam of cesium atoms moves through a vacuum and the atoms are deflected by magnetic fields along their path. The magnitude of the deflection depends on the energy state of the atoms so that atoms in different states are separated into different paths. Zacharias arranged his apparatus to block out all atoms in unwanted states. Only those atoms in one particular energy state were allowed to move into a microwave field set at the particular frequency that induced a transition of atoms in the desired state. Then only these atoms reach the detector. When the applied microwave field had a frequency *exactly* equal to the transition frequency, a maximum number of atoms reach the detector. How does this constitute a clock? A special electronic circuit monitors the number of atoms reaching the detector, and constantly adjusts the frequency of the applied microwave field so as to equal the cesium transition frequency and maintain a maximum number of atoms at the detector. With a maximum number of beam atoms incident on the detector it is known that the applied frequency is exactly equal to the transition frequency. This frequency becomes the "tick" of the clock.

About forty years after Zacharias's attempts, the atomic cesium fountain clock became a reality. Contemporary cesium clocks keep time with great accuracy. The best cesium clocks are fountain clocks and are accurate to about one second in 20 million years. These clocks keep better time than either the daily rotation of the Earth or the annual revolution of the Earth around the Sun. For this reason, a new definition of the basic unit of time, the second, was adopted in 1967. The second, once defined as 1/86,400 of a day, is now defined as 9,192,631,770 periods of the resonance frequency of the Cs^{133} atom. Cesium clocks are commercially available and widely used.

The drive for accuracy produced Ramsey's separated oscillatory field method; however, there were limits to how far apart the two oscillatory fields could be placed without the gravitational force of the Earth pulling beam particles out of the beam. So Ramsey came up with another idea, which led directly to the hydrogen maser clock. The challenge to keep beam atoms in the oscillatory field for a longer period of time prompted Ramsey to wonder if beam atoms could be trapped in some kind of container placed within the oscillatory field and allowed to bounce around for a while in the container before they exited the critical oscillatory field. Why not? In the late 1950s, Ramsey shared his thoughts with Daniel Kleppner, one of Ramsey's most outstanding graduate students who later became a professor at MIT. (Over his career, Ramsey had eighty-four students who received their Ph.D.'s under his direction, and many of them became ground-breaking physicists.) Together, Ramsey and Kleppner invented the hydrogen maser in 1960 and the clock based upon it (see Figure 18.1).[4]

The cesium clock works by detecting atoms at a detector; the hydrogen maser clock uses the direct radiation emitted by hydrogen atoms. In the hydrogen maser clock, hydrogen molecules are dissociated into atoms by a discharge and these atoms diffuse through collimating devices to form a beam of hydrogen atoms. The hydrogen atoms next enter a magnetic field configured to select a particular hyperfine energy state of the hydrogen atom. As an aside, the particular energy state selected by Ramsey and Kleppner is the higher of the same two states that were the subject of Rabi's talk at Shelter Island in 1947 and which led to the correct magnetic moment of the electron. Interestingly enough, these are also the same two states that Edward Purcell and Doc Ewen used to detect hydrogen atoms in space, which were subsequently used to map the Milky Way Galaxy. Ramsey notes: "Dan Kleppner and I invented the atomic hydrogen maser primarily to measure [the energy separation between these two states] more accurately."[5]

In Ramsey and Kleppner's design of the hydrogen maser clock,

Figure 18.1 Norman Ramsey *(center)*, Daniel Kleppner *(right)*, and another Ramsey graduate student, Stuart B. Crampton, standing beside a version of the hydrogen maser clock.

the selected state, the hyperfine state labeled $F = 1$ in the lower right-hand corner of Figure 16.1, is directed into a teflon-coated spherical storage bulb. The teflon coating allows the hydrogen atoms to remain in the higher energy state through some 10,000 collisions with the walls of the storage bulb. All of these hydrogen atoms are poised in their high energy state waiting to decay into the lower state. When one hydrogen atom decays, it emits radiation at a frequency of 1,420,405,721.68 cycles per second. This microwave photon stimulates other hydrogen atoms to decay and a microwave field is soon built up in the storage bulb. This is the maser action from which the device derives its name. (The word maser is an acronym: Microwave Amplification by Stimulated Emission of Radiation.)

The microwave field of frequency 1,420 megacycles per second (Mc/s) is sustained in the storage bulb by the constant entry of hydrogen atoms into the bulb from the incident beam. A tiny pickup probe is inserted into the storage bulb and an electrical current is induced in this probe at the same microwave frequency. This signal is fed into a series of electronic circuits that convert the frequency into timing pulses, or the "ticks" of the hydrogen maser clock.

The hydrogen maser clock is stable to about one second in 300 million years. For many years it has been the most stable clock available. However, there is a drawback to the hydrogen maser clock: its accuracy can only be maintained over a period of a few days. After a few days the stability of the clock deteriorates due to the collisions of the hydrogen atoms with the walls of the storage cavity that alter the resonance characteristic of the cavity and the clock's accuracy is compromised.

Hydrogen maser clocks have been used in many practical applications. For example, they are used to track space probes traveling in the Sun's planetary system and to locate stars or quasars billions of light years from Earth. To accomplish this, two widely sepa-

Figure 18.2 The early hydrogen maser, forerunner of the hydrogen maser clock.

rated radio telescopes, synchronized by hydrogen maser clocks, are used. A star or space probe is detected from the two telescopes. By recording the precise times that similar radiations arrive from the star or space probe at the two separated telescopes, its exact location can be determined. A hydrogen maser has also been used by Ramsey's collaborator Robert Vessot in a high-altitude experiment to confirm that time speeds up as the gravitational force weakens, predicted by the general theory of relativity, to an accuracy of 0.007 percent.[6] Experiments are being designed whose objective is to detect the much-sought gravity waves, another prediction of general relativity.

Closer to home, hydrogen maser clocks have been used to track the motion of the Earth's tectonic plates. From tracking stations separated by 209 miles along the San Andreas fault, it was learned that over an eleven-week period, the distance between the two stations increased by eight inches. An earthquake occurred shortly after this determination. There is a likely correlation between this plate motion and the earthquake. The east-west dimension of the United States is more stable. Tracking stations in Massachusetts and California remained a relatively constant 154,680,381 ± 1 inches apart over a five-year period.

The hydrogen atom has enabled physicists to measure time with unprecedented accuracy, but it has not brought us any closer to an understanding of what time really *is*. Perhaps physics will never unlock this mystery. Einstein recognized that for humans the present means something very different than the past and the future. This meaning, Einstein concluded, cannot be found in physics.

To humans, there is a direction to time: from past to future. A movie showing a chick crawling in among the parts of an eggshell, the shell parts assembling themselves around the chick, and finally the completed egg closing itself and encapsulating the chick would obviously be a movie run backward, thereby running contrary to our sensibilities. At the level of the basic particles of nature, time can run either forward or backward. To say it better, a physicist could not now look at a film of interacting particles and determine whether the movie is running forward or backward. This may be changing, however.

As the twentieth century came to a close, a new puzzle presented itself in the form of experimental results. At the CERN and Fermi laboratories an experiment with kaons and antikaons revealed new insights into the mysterious concept of time. In this experiment kaons and antikaons were created and the oscillation rates of kaons into antikaons and vice-versa were measured. If there was no distinction between time running forward or backward, the two rates would be equal. This was not found. The experimental results showed that antikaons turn into kaons more rapidly than the converse. This result may lead to an understanding of why the universe is dominated by matter rather than antimatter.

If physicists ever think they understand how time really works, it is safe to assume that to subject their suppositions to experiment they will be required to measure time to a high degree of accuracy. If this comes about, the hydrogen atom with its 1,420 Mc/s transition may well be called on to demystify the enigma called time.

19

The Rydberg Constant: A Fundamental Constant

Johannes Robert Rydberg, 1890 •
Theodor W. Hänsch, 1992

The Rydberg constant is one of the most important constants of atomic
physics because of its connection with the fundamental atomic constants $(e,$
$h, m_e, c)$ and because of the high accuracy with which it can be determined.

—George W. Series

The Rydberg constant first appeared in the literature of physics in
1890. Today, more than a century later, this constant still challenges physicists as they carefully design experiments with state-
of-the-art instruments to measure the Rydberg constant with
ever-increasing precision. There are good reasons for the interest
in this constant, but before we consider these reasons, three background questions assert themselves: first, what makes a constant
"fundamental"? Second, where do fundamental constants come
from? And third, why are fundamental constants important?

Constants deemed fundamental are those that emerge from the
core of the overarching theories of physics; they are constants
whose values determine the magnitudes of the basic interactions
of nature; and finally, they are the constants whose values are
linked to and help establish the values of other significant physical
constants. The melting point of water stands alone as an important property of water, but its import does not extend beyond water. The speed of sound is different for every medium in which

the mechanical disturbance called sound propagates. Neither the melting point of water nor the speed of sound is a fundamental constant. By contrast, the fine-structure constant appears in many different physical contexts and links important domains of physics. The fine-structure constant is a fundamental constant.

The fundamental constants originate from both nature itself and physical theories. Every atom in the universe consists of electrons and protons. The charge carried by these fundamental particles as well as their masses are in any list of fundamental constants. A basic property of both the electron and the proton is their magnetic moments. These basic particle properties are fundamental constants that come from nature. The great theories of physics such as gravitation, electromagnetism, relativity, quantum mechanics, and quantum electrodynamics (QED) also give rise to the fundamental constants such as the gravitational constant, G, the speed of light, c, Planck's constant, h, and the fine-structure constant, α.

The importance of the fundamental constants is manifold. The fundamental constants often come into play when experimental results are compared with theoretical predictions. When these constants are known with high precision, they can expose both the strengths and the weaknesses in the theories that physicists employ to explain the physical processes of nature. It is through the interplay between measured result and predicted result that physical theories are put to the test: questioned, refined, or discarded. The lure of measuring the fundamental constants to ever-increasing precision has stimulated new experimental techniques that have paid dividends throughout science. The fundamental constants not only link experiment and theory, they also link different theories. Their appearance in diverse theoretical contexts speaks for their significance.

As another indicator of their basic nature and hence their importance, the fundamental constants often appear grouped together in clusters, so that they provide a natural means of cross-checking one against the other. For example, the ratio of two fundamental constants, *e/h*, consisting of the charge of the electron, *e*, and Planck's constant, *h*, has significance in condensed matter physics, QED, high-energy physics, atomic physics, and X-ray physics. Still another reason underlying the importance of fundamental constants is this: The system of weights and measures has historically been based on artifacts such as the platinum-iridium bar, safeguarded by the International Bureau of Weights and Measures in Paris, which is the standard for length, the meter, defined as the distance between two scratches on the bar. The fundamental constants provide a potential for defining standards on a basis that draws directly from the workings of nature itself and is thereby independent of arbitrary scratches on a metal bar.

The Rydberg constant is a constant that meets all the criteria established for being deemed fundamental. In addition, the Rydberg constant connects all theoretical calculations and experimental measurements of the energy states of any atom.

Fourteen years after Balmer published his work that showed how the wavelengths of the hydrogen spectrum could be represented by a simple mathematical formula, Johannes Robert Rydberg came across Balmer's paper. Rydberg was a mathematician and physicist from Sweden who was fascinated by the periodic system of the chemical elements. Rydberg believed that the spectra of the elements held the key to a deeper understanding of the elements themselves as well as the reasons for the periodicity the elements exhibited in the Periodic Table. He was particularly interested in the observation that the spectra of many chemical elements appear to be members of a series of wavelengths, hinting at relationships among the wavelengths. Rydberg succeeded in developing a simple mathematical formula that accurately reproduced the observations associated with these spectral series.

Just about the time Rydberg composed his simple mathematical formula, he also discovered Balmer's paper on the spectrum of the hydrogen atom. Rydberg realized immediately that he could

recast the Balmer formula for the hydrogen spectrum into the
more general form of his just-discovered mathematical formula.
In Rydberg's recast form, the Balmer formula looked like this:

$$\frac{1}{\lambda} = N\left(\frac{1}{4} - \frac{1}{m^2}\right)$$

where λ is the wavelength of a spectral line, m is a running integer,
and N is a constant, as determined by Rydberg himself, equal to
109,721.6 cm^{-1}. Rydberg claimed that N was a "universal constant
common to all the series and to all the elements examined."[1] With
this formula, Rydberg was able to give an accurate account of the
spectra of many different chemical elements. The Rydberg constant, N, was later given the symbol R in honor of Rydberg. Still
later, the constant was given the symbol R_∞, which reflects the
need to take into account the mass of the nucleus of the atom producing the spectral transitions.

On the basis of Rydberg's work alone, the constant did not
qualify as a fundamental constant. In 1890, when Rydberg published his results, the constant was an empirical number, which
means simply that it emerged from an analysis of experimental
data. No deeper significance could be brought to this number
other than that it was consistent with a good range of physical
data and was therefore probably important.

The status of the constant changed dramatically when Niels
Bohr crafted his model of the hydrogen atom. Specifically, Bohr's
theory revealed that the Rydberg constant was not just a number,
but was a combination of other fundamental constants. Here is
the result that emerged from Bohr's work:

$$R = \frac{2\pi^2 m_e e^4}{h^3 c}.$$

Bohr was able to express Rydberg's constant in terms of the electron's mass, m_e, the electron's charge, e, Planck's constant, h, and

the speed of light, c. After Bohr finished with it, Rydberg's constant was no longer a simple empirical constant, but had a theoretical basis and a deep significance because it was directly linked to other fundamental constants.

The constants that make up the Rydberg constant reach into other major domains of physics. The constants m_e and e are key properties of the electron and each of these constants appears individually or in combination with other constants in many physical contexts. The constant c arises in electromagnetic phenomena and relativity theory, and the constant h is ubiquitous in quantum mechanics. Perhaps more significant, the constants that come together to define the Rydberg constant are, on an individual basis, difficult to measure with great precision. As we shall see, the Rydberg constant can be measured very precisely and thus it becomes one of the cornerstones for the determination of other basic constants. In addition, a highly accurate value of the Rydberg constant would provide a stringent test of QED.

Over the decades of the twentieth century, experimental physicists worked diligently to measure the Rydberg constant with ever greater precision. It has been the hydrogen atom that has provided the principal means for this quest. More specifically, it has been primarily the first and most prominent spectral line in the Balmer series, the bright red H_α line, that has been the focus of interest. For this reason among others, the H_α line "has been studied more intensively than any other line in experimental spectroscopy."[2]

When one considers the myriad of measurements that are made in the conduct of science, the measurements of spectral wavelengths stand apart in that they can be measured with great precision. Thus, from the measurement of the wavelength of the H_α line, physicists over the years have determined the Rydberg constant. A few of the early results are given in Table 19.1.

The experimental results shown in the table cluster around the value 109,737.3 cm^{-1} with uncertainties precluding values

Table 19.1   Values of the Rydberg constant determined by optical
             spectroscopy

Physicist	Year	Value (cm^{-1})
J. R. Rydberg	1890	109,721.6
W. V. Houston[a]	1927	109,737.424 ± 0.020
R. T. Birge[b]	1941	109,737.303 ± 0.017
J. W. M. DuMond et al.[c]	1953	109,737.309 ± 0.012
B. P. Kibble et al.[d]	1974	109,737.326 ± 0.008

a. William V. Houston, "A Spectroscopic Determination of e/m," *Physical Review 30*, 608–613 (1927).

b. R. T. Birge, "The Values of R and of e/m, from the Spectra of H, D and He$^+$," *Physical Review 60*, 766–785 (1941).

c. Jesse W. M. DuMond and E. Richard Cohen, "Least-Squares Adjustment of the Atomic Constants," *Reviews of Modern Physics 25*, 691–708 (1953).

d. B. P. Kibble, W. R. C. Rowley, R. E. Shawyer, and G. W. Series, "An Experimental Determination of the Rydberg Constant," *Journal of Physics B 6*, 1079–1089 (1973).

with greater precision. The situation changed around 1985. The breakthrough that allowed a significant improvement in the measured precision of the Rydberg constant came when physicists confronted an inherent limitation in the measurement of spectral wavelengths. The limitation arises because the H$_\alpha$ spectral transition, as well as all other spectral transitions, does not consist of a single wavelength but a cluster of wavelengths distributed symmetrically around the "true" wavelength. This clustering of wavelengths arises because the hydrogen atoms emitting the detected photons are in random motion. The photons emitted by atoms moving toward the detector are seen by the detector as shifted toward a shorter wavelength—the faster the atom is moving, the bigger the shift. Alternately, the photons emitted by atoms moving away from the detector are seen by the detector as shifted toward longer wavelengths—the faster the atom is moving, the bigger the shift. This shift in wavelength due to the relative motion between the emitting atom and the detector is called the Doppler effect—this same effect is well known in cosmology as the red shift of distant galaxies, which occurs as the universe expands. The challenge facing experimentalists in their quest to measure the Rydberg constant more accurately was to reduce the Doppler effect, to narrow the clustering of wavelengths, to locate the exact center of the cluster of the observed wavelengths more accurately, and thereby to reduce the uncertainties.

The means to accomplish these ends was one rather obvious and one not-so-obvious step. The obvious step was to find a way to lower the temperature of the hydrogen sample, which would decrease the speed of the randomly moving atoms. The lower temperature, however, produced relatively minor improvements. The not-so-obvious step was the ingenious application of lasers to achieve Doppler-free results.

It can be said with confidence that lasers revitalized, if not re-suscitated, studies of the atom. Atomic physics was one of the most active areas of physical research following the creation of quantum mechanics. Already in the 1930s, however, nuclear physics was rising to prominence and after World War II, particle physics and a little later condensed matter physics became dominant areas of interest. The laser provided a new tool for precision studies of atoms, spurring a genuine renaissance in atomic physics about 1970. Precision measurements of the Rydberg constant were a part of this resurgence of physicists' interest in atoms.

Many outstanding physicists have struggled to extend the measured value of the Rydberg constant to the next level of accuracy. Theodor W. Hänsch, however, is one of the most persistent and most successful in designing new experiments to probe the hydrogen atom and establish more accurate values of the Rydberg constant.

Hänsch was born and raised in Heidelberg, Germany. He became interested in hydrogen in 1967 as a graduate student when he heard an inspiring talk by G. W. Series on the hydrogen atom.

Series was known for his extensive work on the hydrogen atom and for his 1957 book, *The Spectrum of Atomic Hydrogen*.[3] From his first paper on hydrogen in 1972,[4] Hänsch's interest in hydrogen has been a staple of his physical research. After spending sixteen years on the faculty at Stanford University, Hänsch returned to his native Germany where he has a dual appointment as professor of physics at the University of Munich and director of the Max-Planck-Institut für Quantenoptik.

Before 1974, all wavelength measurements of the hydrogen H_α line, from which the value of the Rydberg constant was deduced, suffered from Doppler broadening of the spectral line, thereby limiting the accuracy of the result. During the early 1970s, an extremely clever method was devised by Hänsch to eliminate this broadening effect. First, Hänsch developed a new type of dye laser that would be suitable for exciting hydrogen atoms and observing the H_α transition.[5] With this laser, two beams were directed at the sample of hydrogen atoms—one strong beam passed through the sample of hydrogen atoms in one direction and the second, weak beam passed through from the opposite direction. Atoms moving either toward or away from the oncoming laser beams "see" different wavelengths because of the Doppler effect. Only when the wavelength of the two laser beams is equal to the actual wavelength of the hydrogen spectral transition, H_α, do both beams interact with the same group of atoms; namely, those atoms that are effectively standing still.

Here then is the crux of the experiment: in response to the strong beam, tuned to the transition wavelength, essentially *all* hydrogen atoms that are motionless relative to the strong laser beam absorb energy from the beam and make the H_α transition. Essentially, the strong beam clears a path for the weaker laser beam. The weaker beam, passing through the sample in the opposite direction, then passes through and, with no atoms to stimulate, no absorption occurs and the beam exits the sample with essentially the same energy it had on entering. It is the intensity of the weak laser beam that is monitored and the magic wavelength is identified as that particular wavelength that permits the weak probe beam to pass through the sample with its intensity unchanged. This experiment, called saturation spectroscopy, is beautifully conceived and generates excellent results.

This experimental measurement of the H_α wavelength produced a value of the Rydberg constant with an accuracy that was a tenfold improvement over previous experiments. The value of the Rydberg constant that Hänsch and his collaborators obtained was

$$R_\infty = 109,737.3143 \pm 0.001.[6]$$

Two years later, in 1976, Hänsch and a graduate student, Carl Wieman, devised another experimental method, called polarization spectroscopy, that improved still further the accuracy of the Rydberg constant. In this method polarized light is used. Two beams of laser light—one a strong beam that is circularly polarized, the other a weak beam that is linearly polarized—pass through a sample of hydrogen atoms in opposite directions. Once again, only when the wavelengths of the two beams are exactly equal to the wavelength of the H_α transition do the two beams interact with the same class of hydrogen atoms—those that are not moving relative to the light beam. The strong, circularly polarized beam interacts with a select population of hydrogen atoms and essentially removes them from the sample. The linearly polarized weak probe beam, encountering a sample of atoms with a select population missing, has its polarization axis rotated by the remaining atoms and this enables light from the probe beam to be detected. When light is detected, the wavelengths of the two laser beams exactly equals the H_α transition wavelength. From the wavelength, the value of the Rydberg constant can be determined. From this experiment, the determined value was

$$R_\infty = 109,737.31476 \pm 0.00032,$$

which is about three times more accurate than the 1974 result.[7]

During the 1974 and 1976 experiments, Hänsch's attention was shifting from the H_α line of hydrogen to a transition that was long recognized "as one of the most intriguing transitions to be studied by Doppler-free high-resolution laser spectroscopy."[8] This intriguing transition is not a part of the Balmer series. The H_α transition is between the second energy state of hydrogen and the third energy state; that is, from $n = 2$ to $n = 3$. The intriguing transition, part of the Lyman series and called the Lyman-alpha transition, is between the first and second energy states, identified as the $1S$–$2S$ transition: $n = 1 \rightarrow n = 2$. Its fascination arises because this transition is inherently narrow, which means that with Doppler-free techniques, even more accurate values of the Rydberg constant might be obtained.

The Balmer series has wavelengths that are visible; by contrast, the wavelength of the $1S$–$2S$ transition is invisible; it is in the ultraviolet region. Dye laser sources at such short wavelengths did not exist. Once again, an ingenious approach, called two-photon spectroscopy, was devised to open the door to a study of the $1S$–$2S$ transition of the hydrogen atom. In two-photon spectroscopy, the laser beam has a frequency that is half the frequency (or twice the wavelength) of the desired transition. A beam from the laser is sent through the sample of hydrogen atoms, where it strikes a mirror and is reflected back through the sample. Thus, the hydrogen atoms are bathed in laser light, tuned at exactly twice the wavelength of the Lyman-alpha transition, going in opposite directions. Regardless of how a particular hydrogen atom is moving, the Doppler shifts cancel out when an atom absorbs a photon from each of the oppositely moving laser beams. The absorption of the two photons stimulates the Lyman-alpha transition.

Employing the two-photon method, the Rydberg constant

was again measured. From this experiment its determined value was

$$R_\infty = 109{,}737.31492 \pm 0.00022.\text{[9]}$$

Later, with refinements, the two-photon method yielded a more accurate value:

$$R_\infty = 109{,}737.3156841 \pm 0.0000042.\text{[10]}$$

With this result, Hänsch and his collaborators announced in 1992 that "our new value represents the most accurate measurement of any fundamental constant."[11] However, Hänsch's superlative claim had a rather short life.

By October 1997, Hänsch and his co-workers had a new "most accurate" value. Still using the provocative $1S$–$2S$ transition and building on the two-photon method used to get the 1992 result, the 1997 result was

$$R_\infty = 109{,}737.31568639 \pm 0.00000091.\text{[12]}$$

Table 19.2 provides a more complete account of the measured values of the Rydberg constant, including Hänsch's 1997 result.

What is the payoff for this relentless push toward greater and greater accuracy? Why were Hänsch's 1997 results so crucial? The first paragraph of Hänsch and colleagues' 1997 paper explains:

> For almost three decades, the $1S$–$2S$ two-photon transition in atomic hydrogen with its natural linewidth of only 1.3 Hz has inspired advances in high-resolution spectroscopy and optical frequency metrology. This resonance [the $1S$–$2S$ transition] has become a *de facto* optical frequency standard. More importantly, it is providing a cornerstone for the determination of fundamental constants and for stringent tests of quantum electrodynamic theory. In the future, it may unveil

Table 19.2 Measured values of the Rydberg constant

Year	Author	Value
1890	Rydberg	109,675
Later	Rydberg	109,674.7
1914	Curtis	109,737.7
1921	Birge	109,736.9
1929	Birge	109,737.42
1952	Cohen	109,737.311(7)
1959	Martin	109,737.312(8)
1968	Csillag	109,737.3060(60)
1971	Masui	109,737.3188(45)
1973	Kessler	109,737.3208(85)
1973	Kibble*	109,737.3253(77)
1973	Cohen*	109,737.3177(83)
1974	Hänsch*	109,737.3143(10)
1976	Hänsch*	109,737.31476(32)
1978	Goldsmith*	109,737.31506(32)
1980	Petley*	109,737.31529(85)
1981	Amin*	109,737.31544(10)
1986	Hänsch*	109,737.31492(22)
1986	Zhao*	109,737.31569(7)
1986	Hildum*	109,737.31492(22)
1986	Barr*	109,737.3150(11)
1986	Biraben*	109,737.31569(6)
1987	Zhao*	109,737.31573(3)
1987	Beausoleil*	109,737.31571(7)
1987	Boshier*	109,737.31573(5)
1992	Hänsch*	109,737.3156841(42)
1997	Hänsch*	109,737.31568639(91)

* Other individuals were included as authors in the papers reporting the results. Note that in the earlier era it was more common for individuals to work alone.

conceivable slow changes of fundamental constants or even differences between matter and antimatter.[13]

Clearly, these experimental results are not going to impact the lives of today's world citizens. However, from the perspective of the physicist, there is an enormous payoff.

The experiments themselves—saturation spectroscopy, polarization spectroscopy, and two-photon spectroscopy—were magnificent in their design and execution. These experimental methods, developed for the explicit purpose of measuring features of the hydrogen spectrum, will have applications elsewhere. For example, the challenge of measuring the wavelength of the H_α transition with great accuracy motivated the advancement and refinement of lasers and laser techniques, which have wide-ranging applications.

For physics itself, however, the story is still unfolding. When Lamb discovered that the $2S_{1/2}$ and $2P_{1/2}$ states of hydrogen were slightly different, the result became a challenge for theorists, and QED emerged in a refined form as perhaps the most powerful theory of physics. In the experiments of Hänsch, the Lamb shift for the $1S$ ground state of hydrogen was measured with great accuracy. This opens further challenges for physical theorists: the Lamb shift embraces such basic phenomena as the difference between the electron's self-energy in free and bound states, the effect of vacuum polarization on binding energy, and nuclear size effects. A comparison of spectra of the two hydrogens, hydrogen and deuterium, can provide stringent tests to QED, the proton-electron mass ratio (m_p/m_e), and the charge radius of the proton. In fact, Hänsch has already determined the difference between the mean square charge radii of the proton and the deuteron:

$$r_d^2 - r_p^2 = 0.0000000000000038212$$
$$\pm\ 0.000000000000000015\ \text{m}^2.[14]$$

Hänsch has also determined the deuteron structure radius to be

$$r_{str} = 0.00000000000000197535$$
$$\pm\ 0.0000000000000000085\ \text{m}.[15]$$

Since the deuteron is the most important nucleus for understanding the inner workings of nuclei in general, this result will attract the attention of nuclear physicists.

Finally, the Rydberg constant is now known with sufficient accuracy that it may become the basis for a new definition of the second. In this capacity, the Rydberg constant will assume a more visible place in the hierarchy of fundamental constants. Whatever happens, physicists will be scratching their heads trying to devise a way to bring this constant to the next level of accuracy.

20

The Abundance of Deuterium: A Check on Big Bang Cosmology

David N. Schramm, 1945–1997

The last parameter of big bang nucleosynthesis . . . is being pinned down by measurements of the deuterium abundance in high-redshift hydrogen clouds.

—David N. Schramm and Michael S. Turner

In the era of the Hubble Space Telescope it is a common experience to be both intellectually and emotionally stunned by the detailed beauty of images from space. Cosmic clouds, light years in their dimensions, are observed where fetal stars are moving toward birth and delicate nebula mark the graves of deceased stars. No words can capture the awe brought by these cosmic images. It is also commonplace to learn about a new discovery that astronomers and astrophysicists have made. Sometimes, these discoveries are mysterious and take time to embrace with understanding. This was the situation in 1996.

In that year, a group from the California Institute of Technology was engaged in a survey of the sky called the Digital Palomar Sky Survey. During this work, astronomers discovered a strange object glimmering in the night's sky. It was neither a star nor a galaxy. Astronomers remained puzzled by this phenomenon for three years—until mid-1999. In June of 1999, a member of the Caltech team appealed to astronomers at a meeting of the American Astronomical Society in Chicago to help establish the identity of this 5- to 7-billion-year-old mystery object. What is it? Finally, in August, one of the two most powerful telescopes in the world was trained on the light emanating from this mysterious object. Into the bore of the telescope at the Keck Observatory in Hawaii came the wavelengths of light emitted from this strange object and in those wavelengths was a distinct pattern of emissions from the hydrogen atom. These spectral wavelengths emanating from hydrogen atoms in this celestial curiosity enabled scientists to identify it as a quasar, a perplexing quasar to be sure, but a quasar nonetheless.

Once again, hydrogen, the atom of the heavens, came to the assistance of anxious astronomers and helped resolve a puzzling observation. It was hydrogen's Balmer series of spectral lines that, in 1963, helped resolve the enigma presented by mysterious quasi-stellar objects that later came to be called quasars. It was hydrogen's 1420 Mc. hyperfine spectral transition that provided detailed knowledge of galactic structures and astrophysical processes. The prominent place of hydrogen in astronomy and astrophysics should not be a surprise: hydrogen was the dominant, and likely the first, atom to emerge from the big bang.

The big bang is now the accepted cosmological explanation for the origin of the universe. It became so for the best scientific reasons: big bang theory was provocatively suggestive. Both experiment and theory could be brought together to check its predictions and consequences. However, before key predictions were either made or confirmed, big bang cosmology had tough competition.

It is always risky to identify the origin of an idea. The basic idea of the Big Bang may be identified with the Russian mathematical physicist, Alexander A. Friedmann, who in 1922, armed with Einstein's general relativistic equations, developed the picture of the universe expanding from a point origin.[1] The timing was wrong, however, and Friedmann's work attracted little attention. It was another Russian, the playful physicist George Gamow, who came to be identified with big bang cosmology.

Gamow was a nuclear physicist. He imagined a link between the early universe and the formation of the elements. In April 1942, the eighth Conference on Theoretical Physics was held in Washington, D.C. The conference, at Gamow's suggestion, was entitled "Problems of Stellar Evolution and Cosmology." In a report of the conference written by Gamow and J. A. Fleming, we read: "It seems, therefore, more plausible that the elements originated in a process of explosive character, which took place at the 'beginning of time' and resulted in the present expansion of the universe."[2] Gamow and his graduate student Ralph Alpher developed these ideas in more detail in "The Origin of Chemical Elements," published in 1948. The paper was authored by Alpher, Hans Bethe, and Gamow. But Bethe had nothing to do with the paper! Here is a demonstration of Gamow's playful nature. When he was about to submit the paper by Alpher and Gamow, he was reminded of alpha (α) and gamma (γ), the first and third letters of the Greek alphabet. Too bad, Gamow must have thought, that beta (β), the second letter in the Greek alphabet, is missing. So, certainly with a chuckle, Gamow added Bethe's name to the paper and sent it for publication. The Alpher, Bethe, Gamow paper came to be known as the $\alpha\beta\gamma$ paper.[3] (Bethe, with his own good sense of humor, got a kick out of Gamow's stunt.)

In the same year that the $\alpha\beta\gamma$ paper was published, another cosmological scheme was proposed to account for the presently observed universe. The architects of this theory were Fred Hoyle, Hermann Bondi, and Thomas Gold from Cambridge University, who developed it in two papers published in 1948. This theory, called the steady state theory, was the polar opposite of the big bang.[4] Whereas the big bang posited a definite beginning and ongoing evolution, the steady state theory was based on what can be called the perfect cosmological principle: the universe looks the same regardless of where or when observations are made. The perfect cosmological principle has great appeal.

Of course, any cosmology must account for facts such as the observed expansion of the universe. In the face of this incontrovertible observation, steady state theorists had to find a way to keep matter from thinning out as the expansion occurred and to maintain a constant density of matter over time. To do this they made a bold and unfounded basic assumption: they assumed that hydrogen atoms are continuously created throughout all of space. To match the observed expansion and to maintain a constant density of matter, about one hydrogen atom must appear in each and every cubic meter of space every 300,000 years.

The two theories, steady state and big bang, vied for adherents and dominance throughout the 1950s and into the 1960s.[5] But the steady state's appealing principle was no match for the incontrovertible facts that soon surfaced. The relentless adjudication between two strongly held scientific and philosophical positions began. One such fact that required assimilation into cosmological models was the discovery of quasars. Quasars, discovered in 1963, are prodigious sources of energy. Quasars gush energy at such a rate that just one quasar contains the energy equivalent of all the stars in a large galaxy. Equally perplexing were the large red shifts measured for quasars, which meant that all quasars reside at enormous distances from Earth. For the steady state theorists, the distribution of quasars was irreconcilable: for quasars to exist only far away and hence long ago absolutely violated the "same yesterday, today, and tomorrow" as well as "same here as elsewhere" principles of their model.

A second fact that confronted cosmologists in the 1960s was the discovery of a radiation that permeated the universe. This radiation, called microwave background radiation, had been predicted in 1948, but the prediction attracted little to no attention.[6] In fact, the confirmation of this prediction, the discovery itself, was made serendipitously in 1965. Arno Penzias and Robert Wilson, two physicists working at the Bell Telephone Laboratories in New Jersey, were trying to determine the source of interference that accompanied satellite communications. Their objective was to eliminate the troubling interference. They used a horn-shaped antenna similar to though larger than the one used by Purcell and Ewen. No matter where in the Milky Way they pointed the receiving antenna, they detected a background hiss that had the same intensity. They tried everything to eliminate the hiss, including cleaning pigeon droppings that had gathered in the horn-shaped antenna. They finally were forced to conclude that the radiation they were observing was cosmological in origin. They had discovered, by accident, microwave background radiation.

It is sometimes curious how scientists do things. Background radiation is a natural consequence of big bang cosmology: the universe originates at a hot point. With a bang, expansion begins, and then slow cooling. The high-energy radiation that accompanies the big bang also cools and eventually its wavelength stretches to its present microwave dimension characteristic of a cosmic temperature of about 2.7K—just above absolute zero. So why didn't scientists actively look for this radiation? Well, in time they were going to do so. Just as Penzias and Wilson made their discovery, another group of physicists was tooling up to probe the skies for this remnant of the big bang. Quasars and microwave background radiation, among other things, could not be accommodated by the steady state theory, whereas they fit gracefully and logically with big bang cosmology. One of these "other things" brings us back to hydrogen.

The intrigue the big bang held for George Gamow was its potential as the means to bring quantitative understanding to the observed abundances of the chemical elements. Gamow thought that the extreme conditions present during the early moments after the explosion might have provided a caldron in which the elements could have originated. Gamow's fascinating idea proved only partially correct. All the elements except three are synthesized in the interiors of stars. The three exceptions, deuterium, helium, and lithium, are indeed a product of the big bang, whereas the heavier elements—oxygen, nitrogen, magnesium, and iron—result from the fusion of hydrogen, helium, and heavier elements in stars. (Elements heavier than iron are created during stellar explosives called nova or supernova.) So although Gamow's idea missed the bull's-eye, it was certainly on target.

In the first chapter of this book, deuterium was identified as having originated moments after the big bang; thus, deuterium is primordial in character. This raises an important question: Can the currently observed amount of deuterium in the universe become another empirical check on big bang cosmology? More specifically, can the nuclear synthesis of the light elements—mostly helium (^{4}He) plus mere traces of deuterium (^{2}H), helium (^{3}He), and lithium (^{7}Li)—which occurred over a brief period soon after the big bang itself, account for their currently observed abundances?

Many scientists have been interested in this question, but it piqued the interest of one astrophysicist in a sustained and intense fashion. This man was David N. Schramm from the University of Chicago (see Figure 20.1). Schramm was literally and figuratively a big man: he was large in physical stature, in his accommodating personality, in his scientific accomplishments, and in his lifestyle. He drove a red Porsche, piloted his own plane, climbed mountains, and wrestled with professional football players from the Chicago Bears. "All I can say," said Stephen Hawking, "is that David was bigger than life in many ways."[7] It was Schramm who, as much as anyone, brought together the physics of elementary particles and the grandeur of cosmology. David was a first-rate scientist, "but," as astronomer Margaret Geller has written, "per-

Figure 20.1 David Schramm.

haps more important in this harsh world, he was an extraordinary person of great generosity and kindness."[8] "I always considered him to be the leader of our field," said Alan Guth, the creator of inflationary cosmology.[9] In his prime, David Schramm died when the plane he was piloting crashed into a Colorado wheat field on December 19, 1997. He was fifty-two.

Schramm came by this interest in the link between particles and cosmology naturally. He pursued his doctorate at the California Institute of Technology under the direction of William A. Fowler—Willy, as he was known. Also at Caltech were Hubert Reeves and his graduate student Jean Audouze. Fowler was interested in all nuclear astrophysics and Reeves and Audouze were interested in the light elements. Schramm put nuclear astrophysics and the light elements together and it became one of his passions.

Schramm wanted to find out if the abundances of the light elements were consistent with big bang cosmology. To answer this question, he would need to refine theoretical predictions based on the tenets of big bang cosmology, design and carry out astronomical experiments to measure the abundances of the four light elements, and compare the results. As we shall see, the results for deuterium are particularly important—deuterium abundance depends on one and only one important parameter: the density of matter.

Predicting the abundances of light nuclei during the brief epoch of big bang nucleosynthesis (BBN), an event that occurred only moments after the big bang, is a beautiful example of how laboratory physics—hard data—can confront an unobservable cosmological event. The prediction of the abundances of light nuclei during BBN draws directly from data gathered in terrestrial laboratories. Here is the stage for the drama: immediately following the big bang, expansion and cooling began. At about 100 seconds after the big bang, the temperature reached 10^{10}K; after 1,000 seconds, the temperature had dropped to 10^9K. Prior to 100 seconds, at temperatures higher than 10^{10}K, free protons and neutrons existed with such high energies that the strong nuclear force could not bring them together to form a nucleus. At temperatures lower than 10^9K, after 1,000 seconds, nuclei are not moving fast enough to break through the repulsive electromagnetic force acting between their positive charges. Thus, the window for the synthesis of nuclei existed during those nine minutes from about 100 seconds to about 1,000 seconds after the big bang while the expanding universe cooled through the 1,000 degrees from 10^{10}K to 10^9K. The proton and neutron have been studied in laboratories over the energy range experienced by these particles during the period of primordial nuclear synthesis. Thus, the dynamics of their behaviors are known. With this knowledge in hand, predictions can be made.

The predicted abundance of the light nuclei coming out of the big bang depends on one unknown: the number of protons and neutrons present during BBN. Or, since the number of protons and neutrons determines the density of matter during BBN, we can say that the predicted abundances depend on the density of matter. How do we know the density of matter 15 billion years ago? Since the total amount of matter has not changed between then and now, it can be determined by careful estimates of all the visible matter in all the known clouds, dust, planets, comets, stars, galaxies, quasars, and so on. Of course, such estimates are not exact, therefore the estimations are extended over a small range surrounding the best estimate. Over this range, the deuterium abundance is predicted to be in the vicinity of three deuterium atoms for every 10,000 hydrogen atoms. This prediction was made in 1991 by Schramm and others.[10] Shortly before his death, Schramm set out to refine this prediction, which resulted in an abundance of primordial deuterium—just under 3.5 atoms for every 10,000 hydrogen atoms.

Is there any source of deuterium other than that produced by the big bang 15 billion years ago? A number of possible sources have been considered, but Schramm himself developed compelling arguments that deny other possible sources.[11] So we have a prediction: there should be about 3.5 deuterium atoms for every

10,000 hydrogen atoms. Schramm was very confident of this result and called it "very robust" because of the way it dovetailed into laboratory measurements.[12]

Are there processes in nature that destroy deuterium? For certain, any deuterium that was part of a cloud that condensed to form a star would be depleted. Stars fuse hydrogen. There are other processes, though rare, that work to take deuterium out of circulation. Their effect would lead to slight reductions in the observed amount of deuterium. So how does the predicted abundance of deuterium compare with its actual abundance when the universe came out of the period of nuclear synthesis 1,000 seconds after the big bang itself?

The obvious way to determine the abundance of deuterium is to look for its spectral lines, emitted from a nearby cloud of hydrogen. However, since the clouds of hydrogen gas in our galaxy have been cycled in and out of stars several times since the big bang, the deuterium content has been depleted and is therefore not a good measure of its abundance just after the big bang. What we require is a pristine, primordial source of hydrogen that will have hydrogen and deuterium present in the same ratio they had after the period of BBN. A very clever experiment has been designed to show this.

Quasars are at great distances from Earth, which means that we see them now as they were when the universe was very young. On its way to Earth, the light from these distant quasars passes through hydrogen clouds that have not condensed into galaxies, and hence these clouds are free of stars. One such cloud has a red shift that places it approximately 14 billion light years away. Fortunately, behind this cloud is quasar Q0014+813, one of the brightest. The light we see from this quasar passed through the hydrogen cloud 14 billion years ago when the universe was still very young. The deuterium nuclei contained in such a pristine cloud would have only one origin: big bang nucleosynthesis.

Looking at the light coming through such clouds, the presence and abundance of deuterium can be determined. This was the beautiful experiment carried out on Mauna Kea in Hawaii with the ten-meter Keck telescope taking in the light of quasar Q0014+813. The experiment was performed by a team from the Universities of Hawaii and Washington on November 11, 1993,[13] and the results were reported in their paper "Deuterium Abundance and Background Radiation Temperature in High-Redshift Primordial Clouds." The result of this experiment? The observed abundance of deuterium in this primordial cloud was in the range of 1.9 to 2.5 deuterium atoms for every 10,000 hydrogen atoms. This initial experiment has been refined and the results from other groups have pushed this result upward. A more recent result reveals that there are about 3.4 deuterium atoms for every 10,000 hydrogen atoms.[14] Clearly, the predictions of BBN are consistent with experimental results. Therefore, the abundance of deuterium provides a good check on big bang cosmology.

In the process, however, another puzzle emerges. The experimental result of approximately three deuterium atoms for every 10,000 hydrogen atoms, if it survives further experimentation and is verified by other observers, may have another profound impact. The amount of deuterium produced just after the big bang depends, as stated above, very sensitively on the total number of protons and neutrons present during those early moments. However, the total number of protons and neutrons today is the same as the total number of protons and neutrons when deuterium was produced. Thus, knowing the primordial abundance of deuterium tells us the number of protons and neutrons *today*. This same number essentially determines the mass of the visible universe. The resulting mass coming from this process falls short of the mass needed to explain the gravitational behavior of galaxies and their halos. In short, there is missing mass.

This missing mass has come to be called dark matter. Since the deuterium abundance of three deuterium atoms for every 10,000 hydrogen atoms is consistent with the observed number of protons and neutrons, the missing matter must be made up of something other than protons and neutrons. In other words, the missing matter, the matter required to explain the gravitational effects observed in galaxies, exists in some unknown form. The nature of this missing matter is a burning question of astrophysics.

The abundances of the light elements are now considered one of the most stringent tests of big bang cosmology. Among the light elements, deuterium abundance plays a particularly significant role because it depends so sensitively on the density of matter. Thus, the heroine of this book, hydrogen, sits in judgment on the efficacy of big bang cosmology as well as being a primary indicator of the density of matter.

21

Antihydrogen: The First Antiatom

The hydrogen atom may hold surprises yet to come. On the other hand, perhaps the greatest surprise would be none at all.

—Theodor W. Hänsch

The time scales of human experience—hours, months, or decades—provide no mental hooks to grasp either the meaning or the significance of the age of the universe: 15 billion years. Even the time contained in a millennium is not easy to grasp. To identify the age of the universe as 15 million millennia doesn't really help. In the face of this conundrum, it is astounding that events occurring some 15 billion years ago have been described theoretically by big bang cosmology and have been validated experimentally by means of direct observations and laboratory-generated data. The abundance of deuterium, quasars, and microwave background radiation are three empirical pods that endow the big bang theory with intellectual plausibility. As a result, big bang cosmology has the support of most contemporary scientists. In sound science, however, a good theoretical framework like big bang cosmology invokes nagging questions and poses inviting puzzles.

For example, where is the antideuterium? More generally, where is the antimatter? At the microscopic level, that is, the level of fundamental particles that make up the Earth, Sun, and Milky Way, all evidence points to a clear symmetry between matter and antimatter; yet, on the macroscopic level we observe an asymmetry—matter dominates the cosmos. Why should this be? Did the big bang produce equal amounts of matter and antimatter and Earth just happens to reside in a portion of the universe where matter holds sway? In our local cosmic environment we know with full confidence that matter dominates. Matter clearly is the stuff of the toasted cheese sandwiches we enjoy. But beyond lunchtime cuisine, the weight of accumulated evidence suggests that the universe consists of matter, *only* ordinary matter. For example, while the Earth is bombarded by cosmic rays consisting of particles from space, individual antiparticles are not observed nor are the simplest composite nuclei such as an antideuteron or an antialpha particle ever seen. If antimatter was present in other parts of the universe, there would be antinuclei and antiatoms in those regions. One might expect that, among the cosmic rays raining in on the Earth, there would be an occasional antiparticle of some kind. Antiparticles have not been observed in cosmic rays. Also, if there were concentrations of antimatter in the universe, one would expect to observe a background of gamma radiation emanating from those localized spatial regions in between the matter and antimatter domains. Such gamma rays would result from the annihilation that occurs when particles and antiparticles meet. However, spatial regions emanating gamma radiation are not observed. So the lack of evidence to the contrary suggests a universe made up of matter.

The absence of antimatter only spurs more questions. Why did the big bang elect to serve up matter rather than antimatter? Or did the big bang create both matter and antimatter, but in slightly different amounts so that when particles and antiparticles annihilated each other, there was a residue of matter remaining that, after 15 billion years of expansion, makes up the current universe? There are plausible answers to these questions, but the answers are laced with speculation and uncertainties. Speculation is not always unproductive, however. Today's provisional answers are likely to be present in tomorrow's facts.

Physicists currently understand the asymmetry between matter and antimatter in terms of reasoning based on a conservation law and a symmetry principle. If physicists were asked to rank physical principles in terms of their importance, symmetry principles and conservation laws would be prominently located at or near the top of the list. Curiously, both one conservation law and one symmetry principle must be violated in order to provide an answer. The violated conservation law is the conservation of baryon number. Baryons are a class of particles and their antiparticles that include the familiar proton and neutron. There are more massive and less familiar baryons as well. Baryons are given the number +1 and antibaryons the number −1. The conservation of baryon number states that any process involving baryons can occur as long as the baryon number stays the same; that is, the total baryon number before the process must equal the total baryon number after the process. If the big bang produced exactly equal amounts of matter and antimatter and if the conservation of baryon number held precisely, the particles of matter and antimatter would annihilate each other and there would be nothing left in the universe except the background radiation—no galaxies, no stars, no planets, no Picasso.

So, the reasoning goes, there must be a slight breakdown in baryon conservation. Actually, a very, very slight breakdown. The relatively massive proton owes its stability to the conservation of baryons. In general, massive particles are energetically unstable. To achieve greater stability they decay into less massive particles. An isolated neutron, for example, has a mean lifetime of 896 seconds before it decays into a proton and an electron. The less-massive proton and electron are energetically more stable that the more massive neutron. The proton could conceivably decay into a positron and a neutral pion, but in so doing, the conservation of baryon number would be violated. But is the proton absolutely stable? Is baryon conservation absolute?

The proton has an experimentally established lifetime on the order of 10^{30} years, which is about 10^{20} times older than the universe itself. Right now experiments are ongoing in which detectors surrounding some 10^{30} protons are patiently waiting a signal that would indicate the decay of a proton. One proton decay would give the magnitude of departure from baryon conservation. One proton decay would open the way for a universe of matter. (The status of baryon conservation is a bit suspect because an external observer cannot distinguish between a black hole of matter with a positive baryon number and a black hole of antimatter with a negative baryon number. This means that baryon conservation cannot be as deeply significant as charge conservation.) Along with a slight breakdown in baryon conservation, a symmetry principle must be violated.

If the laws of physics retain their validity when a system is observed from different points of view, a symmetry principle can be identified. One symmetry principle called parity, P, states that the laws of physics retain their validity whether one looks at a system directly or looks at its mirror image. Look into a mirror: raise your right arm and salute your image. Your image is seen saluting back, but with the left arm. Mirrors exchange handedness: right for left and vice-versa. The parity symmetry says that nature makes no absolute distinction between right and left, that a process and its mirror image are described by the same physical laws. Physicists believed that P symmetry held absolutely until 1957, when Chien-Shiung Wu and her coworkers[1] and, independently, Richard Garwin, Leon Lederman, and Marcel Weinrich[2] showed by different means that P symmetry was violated.

The demise of the parity principle brought down another symmetry principle that goes by the awkward name of charge conjugation. This principle, C symmetry, simply means that the same laws that describe a process involving matter will also describe the same process involving antimatter. In fact, however, the violation of P symmetry showed that physical laws can distinguish between matter and antimatter.

Whereas both P and C symmetries are individually violated by nature, the combination CP invariance—the exchange of particles with their corresponding antiparticles followed by a reflection in a mirror—was thought to hold absolutely until 1964, when J. W. Cronin, Val L. Fitch, and colleagues showed that CP symmetry was violated in the decay of neutral kaons.[3] This was the one and only violation of CP symmetry that had been observed until March 2001, when the decay of neutral B mesons also violated this symmetry. But these two violations mean that CP invariance is not absolute. With this violation of CP symmetry and with the possibility of baryon nonconservation, it is possible to explain why matter dominates the universe. The explanation, however, is speculative and not very satisfying.

The symmetry principles C, P, and CP are not absolute. The symmetry principle CPT, discussed below, is regarded as absolute and is often referred to as the CPT theorem. This is where antihydrogen comes in. The antihydrogen atom exists; it has been observed. The hydrogen atom consists of a positively charged proton surrounded by a negatively charged electron. In the antihydrogen atom, particles are replaced by their antiparticles: the nucleus of antihydrogen consists of an antiproton carrying a negative charge surrounded by an antielectron, called the positron, carrying a positive charge.

The first hint of the existence of antimatter emerged from a puzzling consequence in Paul Dirac's famous paper of 1928 in which he brought together the theories of relativity and quantum mechanics. After several attempts to understand this hint, physicists proposed the existence of a new particle exactly like the electron, except opposite in charge. On August 2, 1932, Carl D. Anderson, working at the California Institute of Technology, photographed the first cosmic ray track that was made by a positron.[4] Now positrons are routinely created in accelerator laboratories where they have been studied in great detail and have been used as both projectiles and targets for wide-ranging studies. The antiproton was discovered twenty-three years later in 1955 at the University of California, Berkeley by Owen Chamberlain and Emilio G. Segrè.[5] The antiproton also can be made in accelerator laboratories and has been studied rather thoroughly. These two experiments verified that two components of antihydrogen exist, therefore antihydrogen itself must exist. The trick will be to make and assemble antihydrogen in sufficient quantities to subject it to critical study.

Why study antihydrogen? Because hydrogen, in its antimatter form, provides the opportunity to test two bedrock principles of physics: CPT symmetry and the equivalence principle. Once again, the hydrogen atom begs the attention of physicists, who will look to it for enlightenment.

The T in CPT symmetry implies that the laws of physics are valid (invariant) for either time moving forward or time moving backward. To apply CPT symmetry, one applies each symmetry in succession: exchange particles with antiparticles (C), exchange left and right (P), and exchange forward-moving time with backward-moving time (T). The same laws of physics are valid both before and after these exchanges. As discussed above, the CP symmetry is violated in two known cases; however, when T is added, the resulting symmetry, CPT, is an exact symmetry of nature. The kaon system that revealed a violation of CP symmetry is found consistent to a high degree of precision (a few parts in 10^{18}) with CPT symmetry.

CPT symmetry is deeply embedded in fundamental physics. One can prove that quantum field theory and special relativity (as we currently understand these theories) respect CPT symmetry. If CPT were to be proven invalid, contemporary physics would be scrambled.

The equivalence principle is the cornerstone of general relativity. This principle states that the weight of any object is proportional to its inertial mass. This is quite amazing. A brick has an inherent property called its inertial mass; the brick's weight, which is in direct proportion to its gravitational mass, is a measure of the gravitational attraction between it and the Earth. The equivalence principle asserts that these two masses, inertial and gravitational, are identical. Does antimatter obey the equivalence principle? Simply put, physicists do not know.

The antihydrogen atom puts both the CPT symmetry principle and the equivalence principle to the most exacting test now conceived. Before considering how this can be done, it is appropriate to consider briefly how the antiproton and the positron can be brought together to form antihydrogen. Dan Kleppner said in 1992 at a workshop in Munich, "in the past six years the creation of antihydrogen has advanced from the totally visionary to the merely very difficult."[6] Since 1992, the "merely very difficult" remains very difficult.

As stated earlier, antihydrogen has been made: first at CERN in Geneva, Switzerland[7] and a short time later at Fermilab in Illinois (see Figure 21.1).[8] However, only a few antiatoms were produced and observed. The challenge now confronting experimentalists is first to make antihydrogen in quantity, next to bring the antiatoms together, and finally to hold them in isolation long enough for spectroscopy.

The most accurately known physical constant is the Rydberg constant, whose high-precision value was determined from the 1*S*–2*S* spectral transition of hydrogen. The same transition for antihydrogen holds the potential to put both CPT and the equivalence principle to the best available test. A direct consequence of the CPT theorem is that particles and their antiparticles have identical masses, identical charges, and identical magnetic moments. It therefore follows that the spectra of hydrogen and anti-

Figure 21.1 The experimental apparatus at Fermilab that produced the first antihydrogen atom.

hydrogen should be identical. The incredible precision achieved in the study of the hydrogen spectrum has elevated quantum electrodynamics to the status of the most successful physical theory. The same precision achieved with antihydrogen could place CPT symmetry on a firmer footing.

Both antiprotons and positrons emerge from the production process with relatively high energies—much too high to come together and form a docile antihydrogen atom. The first step toward an antiatom is to slow down these antiparticles to tepid speeds so that the electrostatic attraction between the antiproton and the positron allows them to embrace each other. After slowing, the next step is to bunch each of the antiparticles and trap them in a suspended cloud isolated from the walls of the apparatus. Then the clouds of antiprotons and positrons must be merged so that they can form antiatoms. All this presents experimental physicists with a daunting challenge, but they will eventually succeed.

Once the antihydrogen atoms are assembled, light from lasers will be directed through the atoms so as to observe the $1S$–$2S$ transition. The stunning result of Ted Hänsch in his measurement of the Rydberg constant for hydrogen will be compared to the value of the Rydberg as observed with antihydrogen. If the energy states of the two atoms are identical and if the same experimental precision can be eventually attained, this constant will have the value

$$R_\infty = 109{,}737.31568639 \pm 0.00000091.$$

If this result is obtained, the CPT theorem will be confirmed; if the Rydberg has a slightly different value, a hallowed symmetry principle underlying basic physics will be thrown open to question.

The spectral analysis of antihydrogen also suggests a way to test

the equivalence principle. More than one method is being considered, but one of the more fascinating involves the seasonal
changes in the gravitational potential of the Sun at the surface of
the Earth. The distance between Sun and Earth varies throughout
the year, leading to a change in the gravitational potential. What
does this have to do with the hydrogen-antihydrogen spectrum?

The link is Einstein's very famous equation, $E = mc^2$. Because
of the mass-energy equivalence, a hydrogen atom's mass includes
a contribution from the atom's energy. An antihydrogen atom in a
high energy state has more energy than it does in a low state. In
turn, an atom in a high energy state is more massive than when it
is in a low energy state. Further, when the atom is in a gravitational field, its gravitational potential energy is greater when the
atom is in a high energy state. One consequence of general relativity is that the wavelength of a photon increases as the gravitational potential decreases. Thus, when an atom in a gravitational
field emits a photon, the wavelength of the emitted photon is decreased in proportion to the potential energy.

Does antimatter respond to gravity in the same way as matter?
Throughout one year, the potential of the Sun at the Earth varies
because of the eccentricity of Earth's orbit around the Sun. If the
positron reacts differently to gravity than does the electron, then
the frequency of the 1S–2S transition for antihydrogen would
exhibit seasonal shifts different from the 1S–2S transition for hydrogen. If the seasonal shifts in wavelength (or frequency) are
identical for both hydrogen and antihydrogen, the equivalence
principle will be confirmed for both matter and antimatter. If the
seasonal shifts are slightly different, it will open the possibility
that matter and antimatter respond differently to gravitation.

Symmetry typically prompts interest in an object and often endows it with beauty. The Gateway Arch in St. Louis, Missouri,
poised on the western bank of the Mississippi River, has both
symmetry and beauty. Its triangular cross-section is large at the

base and gently tapers as it rises to its 660-foot rounded apex, forming a shape known as an inverted brachistochrone with bilateral symmetry.

Symmetry also brings beauty and simplicity to physics. Symmetries are embedded in nature and the physical laws that describe the phenomena of nature have this same symmetry embedded in them. Symmetry, in turn, means that the laws of physics are invariant under the symmetry operation. For example, parity symmetry, P, implied that the same laws describe physical phenomena when right and left are interchanged. A violation of this symmetry was found that destroyed the absolute character of the right-left invariance.

Steven Weinberg has a particularly provocative and charming way to describe the deep significance of the symmetry principles. He invokes the idea of rigidity.[9] Some theoretical equations contain mathematical terms that can be adjusted to fit the data. Such equations have an inherent flexibility. By contrast, other equations are rigid. There are no ways to adjust them—they are logically rigid. The symmetry principles bring to physical theories a logical rigidity that not only endows such equations with a kind of beauty, but also an enormously enhanced credibility. When the equations emerging from a theoretical framework can be written in only one way and when these equations fit experimental data with exactitude, they must be taken seriously. In sum, symmetry brings rigidity and beauty to physics.

The Gateway Arch has a stainless steel skin. As the light changes minute by minute, the arch captures and reflects light in ways that make it a living structure. Seeing the arch from different vantage points under different lighting conditions preserves the basic character of the structure because of its symmetry. In a similar fashion, the symmetry principles bring life and vitality to the physical laws of nature. And it falls on the simple hydrogen atom to provide the most strenuous test of these principles.

22

The Bose-Einstein Condensate for Hydrogen

Satyendranath Bose, 1924 • Albert Einstein, 1925 •
Eric A. Cornell and Carl E. Wieman, 1995 •
Daniel Kleppner and Tom Greytak, 1998

When Thomas Greytak and Daniel Kleppner at MIT started out 22 years
ago to form a Bose-Einstein condensate by cooling and compressing a gas of
hydrogen atoms, they did not realize just how arduous the journey would be.
—Barbara G. Levy

There are various ways that fame comes to a scientist. For Satyendranath Bose it was asking Albert Einstein to run interference for him. Eventually his name was linked with Einstein's in both a statistical method of dealing with quantum particles, called Bose-Einstein statistics, as well as the peculiar state of matter known as the Bose-Einstein condensate. In addition, Bose had a class of particles named after him: the boson. As this example illustrates, Einstein's scientific influence was telling.

In late 1923, Bose, an Indian physicist from Dacca University in East Bengal, submitted a paper to the British journal *Philosophical Magazine*. Six months later, the editors informed Bose that his paper had been rejected. Bose did not give up. In a letter dated June 4, 1924, Bose wrote to Einstein and included a copy of his manuscript. Bose asked for Einstein's opinion of the paper and whether Einstein would "arrange for its publication in the

German journal, *Zeitschrift für Physik*. . . . Though a complete stranger to you," Bose continued, "I do not hesitate in making such a request. Because we are all your pupils though profiting only from your teachings through your writings." Einstein responded decisively. He translated the paper from English into German and submitted it to *Zeitschrift*.[1] The paper was published with a note by Einstein in which he promised to work out the paper's implications in detail.[2]

The details were significant. In July 1924 Einstein read a paper before the Prussian Academy in which he applied the Bose statistical method to an ideal gas and drew an analogy between a quantum gas and a molecular gas. Over the following few months, Einstein wrote what Martin Klein has called "another of his masterful works,"[3] which was published in January 1925.[4] In this paper, Einstein predicted that the particles of an ideal quantum gas could collect together in the lowest energy state and form what is now called a Bose-Einstein condensate. At the time, physicists regarded Einstein's prediction as a curiosity with little or no physical significance.

Einstein never failed to acknowledge Bose as the initiator of quantum statistics. (Two years later, in 1926, quantum statistics was extended independently by Paul Dirac and Enrico Fermi in what is now called Fermi-Dirac statistics.)[5] But Bose was unable to extract from his work the physical significance that Einstein was able to bring to it. With both Bose-Einstein and Fermi-Dirac statistics on the table, it took a few months for physicists to recognize their applications.

The particles that make up the material world all belong to one of two groups: bosons, the social particles that can come together in the same quantum state, and fermions, the antisocial particles, each of which demands a quantum state for itself. The former obey Bose-Einstein statistics and the latter Fermi-Dirac statistics.

There is another distinction between the two types: All bosons have integral spin—1, 2, and so on—whereas all fermions have half-integral spin—$\frac{1}{2}$, $\frac{3}{2}$, $\frac{5}{2}$, and so forth.

Every chemical element displayed in the Periodic Table has distinctive chemical properties because atoms are made up of protons, neutrons, and electrons, which are fermions. The Pauli exclusion principle requires that no two electrons, like all antisocial fermions, can occupy the same quantum state. Thus, electrons bound to nuclei making up atoms exist in an array of shells that allow all the electrons to exist in their own individual quantum state. The shell structures differ from atom to atom, giving each atom its unique chemical and physical properties.

Quarks, electrons, and nitrogen atoms are fermions; photons, alpha particles, and nitrogen molecules are bosons. Bosons are not restricted by the Pauli exclusion principle and many bosons can occupy the same quantum state; in fact, bosons rather like to come together and populate one quantum state. Lasers are possible because photons are bosons.

Einstein predicted a particular behavior of bosons, the Bose-Einstein condensate, in 1925. A Bose-Einstein condensate has great fascination for physicists not only because it is a unique state of matter, but also because it provides a macroscopic view of quantum behavior. Ordinarily atoms are regarded as particles. However, as quantum theory revealed, atoms have both particle and wave properties. As an atom is cooled, its wavelength increases. If these atoms are bosons and if they can be cooled to the point where their wavelengths begin to overlap, they merge their individual identities, enter a single quantum state, become indistinguishable from each other, and "dance in perfect unison."[6] The collection of atoms becomes, in essence, a single atom that can be directly observed: a macroscopic quantum system.

For many years Einstein's prediction was considered as having only theoretical significance. It was not until 1995, seventy years after the prediction, that Eric Cornell and Carl Wieman first produced this curious state of matter with rubidium atoms.[7] For reasons described below, it was long expected that hydrogen atoms would be the first to yield to the exacting conditions required to achieve Bose-Einstein condensation; however, hydrogen proved to be exceedingly stubborn. So it was not hydrogen, but 2,000 rubidium atoms that huddled together at a temperature of 0.0000001K, 100 billionths of a degree above absolute zero, in the first Bose-Einstein condensate. Nonetheless, it was the experience gained with earlier attempts to condense hydrogen that prepared the way for success with rubidium. So, it is still accurate to say that the hydrogen atom led the way.

Daniel Kleppner became interested in the hydrogen atom early. In fact, one might say that hydrogen is in his academic genes. Kleppner was a student of Norman Ramsey, who was a student of I. I. Rabi. Rabi spent much of the 1930s measuring basic properties of hydrogen and Ramsey, who joined Rabi's group in 1937, was an active contributor in the most important work that came out of Rabi's laboratory. In 1960, Ramsey and his graduate student Kleppner developed the hydrogen maser. With that heritage behind him, Kleppner's interest in the simple hydrogen atom was preordained; thus, it is not surprising when Kleppner says, "For me, hydrogen holds an almost mystical attraction."[8]

Sometime in the mid-1970s, Kleppner and Tom Greytak began to think about creating a Bose-Einstein condensate of hydrogen atoms. To accomplish this feat, a gaseous sample of atoms would have to cool to temperatures edging near absolute zero, yet remain a gas. In other words, the sample of gas could neither liquefy nor solidify. It was this requirement that made hydrogen a prime candidate for the elusive Bose-Einstein condensate.

Both the proton and the electron of hydrogen are fermions with a spin of $\frac{1}{2}$. When they combine to form a hydrogen atom, the atom becomes a boson with a spin of 0 (the electron and proton spins opposing each other) or a spin of 1 (the two spins parallel to each other). When two atoms of hydrogen come together, a hydrogen molecule is formed *if* the two electron spins are antiparallel. By contrast, if the electron spins of the two hydrogen atoms are parallel, the two atoms *cannot* combine to form the hydrogen molecule. A group of such hydrogen atoms—called spin-polarized when their electron spins are parallel—can be cooled to absolute zero without forming either a liquid or a solid. The atoms of spin-polarized hydrogen behave as an ideal gas even at the lowest temperatures possible. This was one characteristic of hydrogen that made it such a seductive target for achieving a Bose-Einstein condensate.[9]

Other properties of hydrogen also made it a prime candidate for Bose-Einstein condensation, such as its low mass. The smaller the mass of a particle, the longer its de Broglie wavelength. It was reasoned that hydrogen atoms would not have to be cooled as much as more massive atoms to achieve significant overlapping of their associated wavelengths. It also appeared that methods to cool hydrogen were available and understood.

But hydrogen did not yield easily. Kleppner, Greytak, and their students began work in 1978. They developed methods to trap hydrogen atoms by means of magnetic fields and to cool them by means of evaporation. In evaporation, the faster atoms are allowed to escape the trapping container carrying with them excess energy and leaving behind the cooler atoms. The experimental methods were demanding and intricate. Although these methods were conceived and executed well, they did not succeed with hydrogen. These same methods, however, were the starting point for Cornell and Wieman in 1989 when they decided to apply them to one of the alkali metals, rubidium. With rubidium they successfully created the first Bose-Einstein condensate in 1995.

Kleppner and his students were foiled in their early attempts to realize Bose-Einstein condensation with hydrogen because hydrogen atoms were lost through recombination on the walls of the container. In response, they developed a no-wall container with magnetic fields. They successfully cooled the atoms to around 0.00010K by means of evaporation, but then they hit a temperature wall. Fast hydrogen atoms did not evaporate, as expected, and carry the sample temperature lower. They applied an ingenious technique using radio-frequency (RF) electromagnetic waves. By juggling the frequency of the RF and the magnetic fields they were able to target the most energetic atoms and effectively whisk them away from the sample, thereby leaving the remaining residue colder. This brought the remaining hydrogen atoms down to a temperature of about 0.000050K and at this juncture a twenty-year quest ended: a Bose-Einstein condensate consisting of about 1 billion (10^9) atoms was observed late in the summer of 1998.[10] The number of atoms in the hydrogen Bose-Einstein condensate was much larger (1 billion) than the 2,000 that had been achieved with rubidium, making the hydrogen condensate attractive for further study.

As soon as Bose-Einstein condensates were created, physicists recognized that they were going to be a rich object for study. Theodor Hänsch has said, "It is like a door that has opened to a new world."[11] The new world portends both practical applications and opportunities to extend theoretical understanding.

For example, atoms in a Bose-Einstein condensate are analogous to photons in a laser. The novel features of a laser are achieved because the photons are optically coherent, which means that every photon has the same frequency and phase. The atom waves in a condensate are also coherent. Wolfgang Ketterle and co-workers at MIT beautifully demonstrated the single-wavelength nature and the coherence of the wave nature of atoms in condensates by merging two clouds of condensate atoms and observing interference fringes in the overlapping region.[12] This coherence opened the possibility of atom lasers, which in fact have already been demonstrated,[13] although it is not clear whether atom lasers can be manipulated or sharply focused. One difference between photons and atoms is that photons in a laser beam do not interact with each other. Atoms do interact. What consequences will arise from their interactions? Optical lasers are used for lithography. Physicists are imagining lithographic applications where condensates with the proper atoms are finely focused and used to deposit atoms on a surface to form transistors and other practical devices.

Bose-Einstein condensates are unusual in numerous ways. With careful study physicists will gain basic knowledge about the material and quantum worlds. The atoms in a condensate are indistinguishable. All atoms move at the same speed in the same space. One can ask: How can two objects occupy the same place at the same time? A condensate is a macroscopic quantum wave packet and a macroscopic example of Heisenberg's uncertainty principle. Condensates hold the promise of bringing new insights to the strange world between the microscopic quantum and the macroscopic classical domains.

When atoms congeal into a condensate, they form a very dense medium. The speed with which light propagates in a medium is dependent on, among other things, the density. What is the speed of light in a Bose-Einstein condensate? Lene Hau and her co-workers have answered this question for a condensate of sodium atoms. She directed laser light through a condensate and found that in this curious medium the light crawled through it at the unbelievable rate of seventeen meters per second—some 999,999,983 meters per second less than light moves through a vacuum![14]

In this dense form of matter, condensates will become an arena where the physics of many-body systems can be tested. As a strange mixture of a fluid and a coherent wave, condensates will bring together the high precision of atomic physics with the theoretical frameworks of many-body physics.

In all of the excitement generated by the creation of Bose-Einstein condensates, hydrogen offers its own customary charm. Precision spectroscopy awaits. The frequency of the $1S$–$2S$ transition measured so precisely by Hänsch may be measured with even greater precision. With hydrogen, more atoms can be brought together in a condensate than is the case with other atoms. Hydrogen's simplicity allows theory to be applied to its interactions with exactitude.

All this bodes well for hydrogen's role in elucidating the intricacies of this novel state of matter. Kleppner's "mystical attraction" to hydrogen is certainly justified.

For achieving a Bose-Einstein condensate in dilute gases of alkali atoms, Eric A. Cornell and Carl E. Wieman were awarded the 2001 Nobel Prize in Physics. Sharing the prize with Cornell and Wieman was Wolfgang Ketterle for his fundamental studies of condensates.

23

Exotic Hydrogen-like Atoms:
From Theory to Technology

> Muonic hydrogen, the $\mu^- p^+$ bound state, . . . is more sensitive to the . . . structure of the proton.
>
> —T. Kinoshita and M. Nio

The hydrogen atom has been such a provocative and productive window onto the physical universe that physicists have gone to every length possible to create hydrogen-like systems. Such hydrogen-like "atoms" consist of two particles, one positively charged (like the proton) and the other negatively charged (like the electron), bound together by the mutual attraction of their opposite charges. Just as the hydrogen atom provides an exacting test of physical theories, these quasi-atoms can stretch the demands on theoretical concepts in novel ways. There are three classes of hydrogen-like atoms. The first class goes by the name exotic atoms, the second class are called Rydberg atoms, and the third class are ordinary atoms whose electrons are all stripped away, except one, leaving a lone electron around a highly charged nucleus.

In the case of an exotic atom, physicists bring together an exotic particle and an ordinary particle or two exotic particles to form a short-lived entity that is structurally like hydrogen. The exotic particles involved are unstable and have short lifetimes. For example, one such exotic particle, the muon, has a mean lifetime of 2.1971×10^{-6} seconds; the positron is stable when isolated, but it annihilates quickly when it encounters an ever-present electron in typical laboratory environments; the pion has a mean lifetime of 2.603×10^{-8} seconds. Hydrogen-like atoms constructed from these particles represent challenging objects of study, but the dividends that come with success make the challenge well worth the effort.

In positronium a positron is substituted for the proton and the positron and an electron are bound together in a hydrogen-like atom. Muonic hydrogen, perhaps the simplest exotic atom, unites a negatively charged muon with a positively charged proton. In this instance, a muon simply replaces the electron. In muonium, an electron is bound to a positive muon. As we shall see, there are other examples of exotic atoms. Positronium, muonic hydrogen, and muonium have been studied rather extensively.

The challenge presented by these hydrogen-like atoms is at least twofold. First, the particles must be brought together so that their electrical attraction can bind them into an atom-like unit. Since these exotic particles are formed at energies much too high to allow binding to occur, the particles must be de-energized (slowed down) to the point that their mutual electrical attraction can take hold. Second, experimental techniques must be devised that permit these short-lived hydrogen-like atoms to be studied.

Why go to the trouble? The fascination with hydrogen-like atoms starts with the same fascination physicists have with hydrogen itself. The reason is simplicity—just two particles are involved. For such a simple system, physicists can apply basic physical theories such as quantum mechanics, relativity, and quantum electrodynamics (QED) with minimal assumptions compromising the outcome. Often, for two-particle systems, physicists can solve the mathematical equations that arise exactly. This is not the case with the next simplest atom, helium—a three-particle system—and a relatively simple atom such as carbon confronts physical theory with formidable problems. The fascination is extended because these hydrogen-like atoms provide the opportunity to subject basic theories to new and stringent tests. No matter how successful a theory seems to be, physicists are always eager to expose it to situations where either the theory is further substantiated or it fails. Failure simply raises new questions and holds the potential for important new insights.

Positronium was the first exotic atom to be observed. The positron was predicted from Dirac's 1928 work in which he united the new quantum mechanics with relativity, and it was first observed by Carl Anderson in 1932. Soon thereafter, in 1934, Stjepan Mohorovicic predicted that an "atom," consisting of a positron and an electron, could be observed.[1] It wasn't until 1951, however, that positronium was created and observed by Martin Deutsch of the Massachusetts Institute of Technology.[2] Deutsch accomplished this by slowing positrons emitted in the decay of an isotope of sodium until the positrons captured electrons from the surrounding gas.

Muonic hydrogen was first observed in 1956 by L. W. Alvarez and colleagues.[3] When evidence for the muon first appeared to S. H. Neddermeyer and C. D. Anderson in 1937 of the California Institute of Technology, and was quickly confirmed by J. C. Street and E. G. Stevenson of Harvard, it effectively threw a wrench into the physical works.[4] "Who ordered that?" asked Rabi, with obvious surprise and consternation. For years a three-word sentence adorned the blackboard in Richard Feynman's Caltech office: "Why the muon?" The muon question obviously intruded on Feynman's thoughts; in fact, the same question has troubled the thoughts of many physicists. Muonic hydrogen was first observed serendipitously in an experiment in which particles called K mesons were being stopped in a ten-inch hydrogen bubble chamber. (K mesons are about half as massive as a proton and can carry no charge, positive charge, or negative charge.) In the beam of K mesons, there were also muons, some of which formed a bound system with protons. Muonium itself was discovered by Vernon W. Hughes and his associates in 1960.[5]

These hydrogen-like atoms are, as the name indicates, like hydrogen, but they have crucial differences. In positronium the positron and the electron, each with the same mass, revolve around a point midway between them, which is their common center of mass. (This is just the same as the hydrogen atom except the proton is almost 2,000 times more massive than the electron so that the latter seems to revolve around the former.) From Bohr's model of the hydrogen atom, we can predict that the spatial extent of positronium, in its lowest energy state, is larger than is hydrogen. In fact, it is about twice as large.

The muon is a weighted-down electron. The mass is 206.8 times larger than the electron. When the muon takes the place of the electron in muonic hydrogen, it orbits much closer to the proton so the size of the muonic hydrogen atom is smaller. The radius of the lowest energy state in hydrogen is about 0.5Å whereas in muonic hydrogen it is about 0.003Å. In the muonium atom, a positive muon becomes the effective nucleus. It is much lighter than the proton; specifically, the muon is about one-tenth the mass of the proton. This means that the muonium atom is larger than hydrogen.

The size differences between hydrogen and its counterparts are interesting and important. But there is more. The muon and the positron, like the electron, belong to the particle family called leptons. Leptons are immune to the strong force that acts between nuclear particles—protons and neutrons. It is this strong force that overwhelms the electrical repulsion between protons and brings them together to form an atomic nucleus. Neither the positron nor the muon feel this strong force. This means that exotic atoms are even simpler than hydrogen with its proton nucleus, the source of the strong force.

Vernon Hughes has stated the motivations for studying exotic atoms. These include the opportunities to determine the properties of the particles themselves, to study the interactions among the bound particles, to test modern theory, and to search for the effects of weak, strong, and unknown interactions on the bound state of the particles.[6] As Hughes has said, the study of exotic atoms serves fundamental purposes. In both positronium and muonium, the interacting particles are both leptons; thus, the two particles interact exclusively through the electromagnetic interaction without the strong force hovering in the background. This means that these "atoms" provide an unobstructed view of the electromagnetic interaction with no possible influence, or interference, from the strong force. Or consider muonic hydrogen, in which the muon orbits much closer to the proton and thus opens the possibility of probing structural properties of the proton from a closer vantage point as well as witnessing any influence of the strong force close-up.

The spectrum of hydrogen has been studied more intensively and more exhaustively than any other atom. The energy states of hydrogen are known about as well as anything can be known. As is the case with hydrogen, probing these exotic "atoms" with light holds the possibility of examining their spectra and thereby exposes their physical properties and opens opportunities to test physical theories. In principle, positronium and muonium should permit physicists to make predictions that make demands on theory in addition to those made by hydrogen. For example, unlike the proton, which is made up of quarks, both the positron and the muon are structureless particles, which means that the spacing of their energy states is determined only by the QED interaction. This makes these hydrogen-like atoms an ideal arena for testing QED. Further, the corrections to theory demanded by the Lamb shift in these exotic atoms could, in principle, mean that these theories face new, stringent tests. Quantum transitions among various energy states in both positronium and muonium have been observed and measured. For example, just recently, the 1S–2S energy interval in muonium was measured,[7] and its experimental result was consistent with theory.

The study of exotic atoms is an active area of contemporary research. Many hydrogen-like atoms have been detected and investigated. Among them, in addition to those mentioned above, is pionium, in which an electron is bound to a positive pion and there are hydrogen-like atoms in which the electron has been replaced by exotic particles such as the negative pion, the negative kaon, and the antiproton. As accurate data accumulate, physicists will enjoy searching for the subtle new insights that these curious atoms are likely to provide. Certainly, exotic hydrogen-like atoms will expose the intricacies of physical theory in unique and provocative ways.

The second class of hydrogen-like atoms is called Rydberg atoms. A Rydberg atom is an ordinary atom in which one electron has been elevated to a very high quantum state. The energy states of atoms are identified with the quantum number n, called the principal quantum number. The ground state, or lowest state, is the $n = 1$ state, which is where atoms spend most of their time. The first excited state is the $n = 2$ energy state, the second excited state is $n = 3$, and so on.

The size of an atom is determined by the average distance between the nucleus and the outermost electrons when the atom is in its ground state. When the hydrogen atom is in its ground state, its sole electron is, on average, about 0.5Å away, so the hydrogen atom has a diameter of about 1Å. If the electron in the hydrogen atom is excited to the $n = 2$ energy state, the diameter of the atom increases by a factor of n^2, or by a factor of four. When hydrogen becomes a Rydberg atom, its electron can reside in energy states with $n = 80$, 90, or higher. Rydberg atoms have been observed with n equal to several hundred. This makes Rydberg atoms very large—up to 100,000 times the size of an atom in its lowest quantum state. Like exotic atoms, Rydberg atoms are very fragile, but unlike exotic atoms, they are very long-lived, provided they are isolated and free from collisions by other atoms.

All Rydberg atoms are hydrogen-like. This is because the electron in an elevated energy state is far from both the nucleus and all the other electrons that, in their normal quantum states, remain relatively close to the nucleus. Therefore, if a Rydberg electron could look inward toward its nucleus, it would see a compact sphere consisting of Z positive nuclear protons (plus neutral neutrons) closely surrounded by $Z-1$ negative electrons. Thus the Rydberg electron moves around a core with a net charge of $+1$, just like a hydrogen atom.

Although the existence of Rydberg atoms has been known since the late nineteenth century, the first hydrogen Rydberg atom was probably observed in outer space in 1965.[8] (The qualifier reflects the difficulty of pinning down the first of most anything.) Not only is hydrogen present throughout space, but space provides a congenial environment for Rydberg atoms. Such atoms are fragile and cannot endure the bombardment that would occur in a gaseous sample in a laboratory on Earth. In space, however, each atom is serenely isolated and undisturbed by collisions with other atoms.

Rydberg atoms began to appear in laboratory experiments in the 1970s when tunable dye lasers became a powerful means to create and study them in controlled detail. With these lasers, scientists can excite the outer electron of essentially any atom and form a hydrogen-like Rydberg atom. The most commonly used atoms for Rydberg studies are the alkali metals: lithium, sodium, potassium, and so on. To keep Rydberg atoms from being pummeled by other atoms, destroying their delicate status, experimenters form a beam of atoms moving through a high vacuum. In such beams, atoms move parallel to each other and are effectively captured in isolation.

Apart from the strangeness of these huge atoms, why do scientists study them? What can be learned from Rydberg atoms? One motivation for studying them is their spectacular response to both electric and magnetic fields. When an electric field is applied to an ordinary hydrogen atom, its energy states are slightly shifted. However, when hydrogen becomes a gigantic Rydberg atom, its response to an electric field is likewise gigantic. The energy states shift by large amounts and, in the process, they sometimes cross each other. At the point of crossing the two states have the same energy, that is, they are degenerate. Degenerate energy states are provocative because they imply an underlying simplicity.[9] Discovering such simplicities will bring new insights into the world of atoms.

Another reason for studying Rydberg atoms is that they have the potential for shedding light on the dim domain between the quantum world of the atom and the classical world of everyday objects. Quantum mechanics is the enormously successful theory that describes the building blocks of the physical universe. As such, physicists regard quantum mechanics as more basic than the classical laws of physics. Physicists like to think, and have good reason to do so, that as quantum mechanics is applied to larger and larger objects, quantum physics should be seen to blend into classical physics; that is, in the domain between the microscopic and macroscopic worlds, an equivalence between quantum and classical physics should emerge. This idea of quantum theory merging into classical theory was explicitly expressed by Niels Bohr in his correspondence principle.

The quantum world is characterized by abrupt discontinuity; the classical world by continuity. The low energy states of an atom are distinctly quantum-like: the energy differences among neighboring energy states are large. By contrast, the energy differences among the high energy states of a Rydberg atom are small and, in energy terms, the transitions between them are relatively smooth. From low energy states to high energy states, atoms move from jolting discontinuity toward smooth continuity, from the distinctly quantum domain toward the classical domain.

What would a classical atom look like? Consider the planetary system. A planet is located at a particular position on a well-defined orbit. Contrast that with the ground state of the hydrogen atom, where the electron cannot be located in a particular position nor can a definite orbit be identified. With keen foresight, Heisenberg essentially banished orbits and positions on orbits from his thinking when he created the first version of quantum mechanics. He did this because neither electron orbits nor positions could be either observed or measured and he concluded that immeasurable concepts had no place in a physical theory.

Perhaps it is fortunate that Heisenberg did not know about Rydberg atoms. In recent years, long after Heisenberg's death, physicists have devised very clever ways to examine Rydberg atoms. They have found that in such atoms the electron can be partially localized on an orbit that traces out a distinctly elliptical path.[10] Physicists do not find an electron, but a bell-shaped blob that moves along an elliptical orbit. Like the planets in their orbits, the electron-blob moves most rapidly when it is closest to the center of the Rydberg atom and most slowly when it is farthest. In short, the Rydberg atom exhibits behavior that is consistent with the laws of classical physics.

It is increasingly important to understand nature on a scale in between the quantum and classical worlds. Technological methods have moved into this transitional realm with dramatic results. Scanning tunneling microscopes can image individual atoms and reveal the atomic character of surfaces of solids. Individual atoms can be moved about and materials tailored for specific purposes. Electronic circuit elements have been reduced to dimensions of molecules. As scientists come to understand how the quantum domain gives way to the classical domain, these technologies will multiply and their applications expand. Hydrogen-like atoms may benefit technological development.

The third class of hydrogen-like atoms are highly ionized atoms; that is, atoms whose electrons have been stripped away leaving only one electron in orbit around the nucleus. In hydrogen, the electron moves under the electromagnetic influence of the proton. However, as Willis Lamb's experiment revealed, the hydrogen's electron moves in a space that is teeming with activity: electron-positron pairs pop into brief existence and virtual photons all exert their influence. This lively environment causes the shift in hydrogen's $2S_{1/2}$ state, now called the Lamb shift, and this shift provided a test of QED.

Now consider a highly ionized atom. Uranium is an example. Uranium has ninety-two electrons orbiting around a nucleus with ninety-two protons and a larger number of neutrons. Experimentalists have been able to strip away ninety-one electrons, leaving one electron revolving around the highly charged uranium nucleus, thereby forming the U^{+91} ion. This electron moves in a space much more strongly influenced by electromagnetic effects than is the case in the hydrogen atom. In fact, the sole electron around a uranium nucleus experiences an electric field many times stronger than any field that can be produced in the laboratory. What does the Lamb shift look like for this hydrogen-like atom? Does QED provide an explanation? The answer to the first question is amazing: the Lamb shift is about 1×10^8 times larger in U^{+91} than it is in hydrogen. Now, can QED explain this large shift? The observed Lamb shift is essentially consistent with the prediction of QED.[11] The experimental uncertainties carried by this result are relatively large, but work is underway to refine procedures and increase the precision. Will QED be found wanting? It remains to be seen.

"Imitation," the saying goes, "is the sincerest form of flattery." Hydrogen-like atoms, whether they are exotic atoms comprised of short-lived particles, bloated Rydberg atoms, or highly ion-ized uranium atoms, share the structural features of hydrogen. As such, they are simple and there is nothing more appealing to a physicist than simplicity. Hydrogen and its surrogates continue to reward physicists with the never-ending bounty only the simplest atom can provide.

Epilogue

To understand hydrogen is to understand all of physics.

—Attributed to Victor Weisskopf

When scientists analyze materials distributed within the Earth's crust and throughout the atmosphere, they identify ninety-two individual elements that appear in pure form or in combination with other elements. In addition to the naturally occurring elements, some elements are created in the laboratory, but they are short-lived and not observed in nature. All these chemical elements, both natural and laboratory-made, make up the Periodic Table that hangs in the world's science classrooms. Element fourteen, silicon, is prominent in the sandy beaches that attract summer vacationers, whereas elements one and eight, hydrogen and oxygen, make up the waves that surfers ride until those waves break upon the sandy beach. Hydrogen, carbon, nitrogen, and oxygen, elements one, six, seven, and eight, are the primary building blocks of flowers, squirrels, baseball players, and all other forms of living matter. The Earth is blanketed by a thin layer of gases made up mostly of nitrogen, oxygen, and traces of a few other elements.

The element hydrogen, which has consistently turned new ground in the quest to understand the intricacies of the natural world, is the dominant form of matter in stars and the interstellar regions of the cosmos, but the hydrogen atom is largely absent in pure form here on Earth. There is no hydrogen in the Earth's atmosphere. The gravitational pull that successfully holds nitrogen and oxygen to the Earth's surface cannot maintain its grip on the simplest atom. Atomic hydrogen is too ephemeral. Any hydrogen released to the atmosphere slowly works its way through the nitrogen and oxygen surrounding the Earth until it reaches the upper limits of the Earth's atmosphere, then effectively bids adieu to planet Earth as it disappears into the solar system and beyond. There is a touch of irony that the hydrogen atom, so abundant in the universe at large, yet essentially absent in pure form in the world around us, has brought such insights in return for the attention that it has received.

The story of hydrogen in the preceding account has been drawn from physics. Chemists could add to this tale. One topic that begs for inclusion is hydrogen bonding. This bond is not really a chemical bond such as occurs between the oxygen and the two hydrogen atoms in a single water molecule. The hydrogen bond, however, does occur in water between the oxygen of one molecule and the hydrogen of a neighboring molecule. It is this attraction between oxygen and hydrogen of separate molecules that gives water some of its unusual properties. Ice floats because of the hydrogen bond. And because ice floats, it can be argued that life on Earth exists. If water was a typical liquid, its solid form, ice, would be denser than the liquid form. Ice would then sink to the bottom of bodies of water and slowly build up until the Great Lakes and other bodies of water became solid ice. Since living organisms most likely took form first in water, it is questionable whether solid bodies of ice would have been conducive to the fostering of life. There are other stories that chemists would include in their story of hydrogen.

I believe that the hydrogen atom, the essential element, shows the conduct of science at its best. It was my privilege to write a biography of I. I. Rabi.[1] I spent many days with Rabi and learned firsthand about his work on the hydrogens. This background surely influenced my decision to write a book about the hydrogen atom. This book was born, however, while I was listening to a talk at a meeting of the American Association for the Advancement of Science. The speaker was expounding boldly on current physics and how a single theory of everything was likely to become a reality. "Come on," I said to myself, "the hydrogen atom, the simplest atom, still beckons. We are still learning from the hydrogen atom." After all, H stands not only for hydrogen, but also for humility. The hydrogen atom still beckons—its story far from over. On occasion we hear that all basic knowledge in science has been acquired. Whenever someone makes such claims, it would be advisable to remember that the simplest atom, one proton and one electron, is still providing insights into natural phenomena. As long as scientists are learning from the essential element, hydrogen, science itself is in no danger of ending.

Notes

Prologue

1. Evidence does suggest the presence of "dark" matter, whose nature is yet unknown.

2. John S. Rigden, "H Stands for Hydrogen . . . and Humility," *American Journal of Physics 50*, 299 (1982).

1. In the Beginning

1. For a wonderful account of the early moments of the universe, read Steven Weinberg's *The First Three Minutes: A Modern View of the Origin of the Universe* (New York: Basic Books, 1977).

2. Hydrogen and the Unity of Matter

1. William Prout, *Annals of Philosophy 6*, 321–330 (1815); 7, 111–113 (1816).

2. Quoted in *Prout's Hypothesis*, Alembic Club Reprints, no. 20 (Edinburgh: The Alembic Club, 1932), 17.

3. Ibid., p. 18.

4. Ibid.

5. Frederick Soddy, *Nature 92*, 399–400 (1913).

3. Hydrogen and the Spectra of the Chemical Elements

1. Quoted in Herbert Dingle, *The British Journal for the History of Science 1*, 199–216 (1963), p. 200.

2. Ibid., p. 201.

3. Arthur Schuster, "On Harmonic Ratios in the Spectra of Gases," *Proceedings of the Royal Society 31*, 337–347 (1881).

4. 1 Angstrom = 0.0000000001 meter = 1×10^{-10} meter.

5. Johann Jakob Balmer, *Annalen der Physik und Chemie 25*, 80–85 (1885).

4. The Bohr Model of Hydrogen

1. J. J. Thomson, *Philosophical Magazine 44*, 292–311 (1897).

2. William Thomson, *Popular Lectures and Addresses*, vol. 2 (New York: Macmillan, 1894).

3. Quoted in Abraham Pais, *Inward Bound* (Oxford: Oxford University Press, 1986), p. 179. Thomson's number of 1,000 electrons was possibly prompted by his 1897 data, which showed that the m/e ratio for hydrogen is 1,000 times greater than the m/e ratio for the electron.

4. This quotation appears in numerous references: E. N. da C. Andrade, *Rutherford and the Nature of the Atom* (Garden City: Anchor Books, 1964), p. 114; Barbara Lovett Cline, *The Questioners* (New York: Thomas Y. Crowell, 1965), p. 6; David Wilson, *Rutherford: Simple Genius* (London: Hodder and Stoughton, 1983), p. 296.

5. According to Bohr (oral history interview of Niels Bohr by Thomas Kuhn in 1962), he heard Rutherford lecture in the fall of 1911 during which Rutherford described his new atomic model.

6. Niels Bohr, *Philosophical Magazine 26*, 1–25, 476–502, 857–875 (1913).

7. Niels Bohr, *Collected Works*, L. Rosenfeld, general ed., vol. 2, *Work on Atomic Physics (1912–1917)*, U. Hoyer, ed. (Amsterdam: North Holland, 1981), p. 111.

8. Ibid., p. 123.

9. Quoted in Pais, *Inward Bound*, p. 208.

10. Victor Weisskopf, in *Niels Bohr: Physics and the World* (New York: Harwood Academic Publishers, 1988), p. 4.

5. Relativity Meets the Quantum in the Hydrogen Atom

1. See Werner Heisenberg, *Physics and Philosophy: The Revolution in Modern Science* (New York: Harper & Brothers, 1958), p. 31.

2. Niels Bohr, *Collected Works*, vol. 2, p. 603.

6. The Fine-Structure Constant

1. The dimensionless character of the fine-structure constant can be seen by recognizing that the unit of electronic charge, the coulomb, is not a basic unit; rather, it is a derived unit. In terms of basic units, the coulomb can be expressed as follows: coulomb = newton m. Thus, the fine-structure constant is dimensionless.

2. A muon belongs to the same family as the electron. In a sense, the muon can be regarded as a heavy electron where $m_\mu = 207\ m_e$.

3. See Toichiro Kinoshita, "The Fine Structure Constant," *Reports on Progress in Physics 59*, 1459–1492 (1996).

4. Edward M. Purcell, letter to Mr. Roger Auten, 3 September 1986, in author's private collection.

5. Pais, *Inward Bound*, p. 215.

6. Max Delbrück, "Out of This World," in *Cosmology, Fusion & Other Matters*, Frederick Reines, ed. (Boulder: Colorado Associated University Press, 1972), p. 283.

7. The Birth of Quantum Mechanics

1. Pais, *Inward Bound*, p. 134.

2. Ibid., p. 252.

3. Ibid., p. 248.

4. Quoted in Jagdish Mehra and Helmut Rechenberg, *The Historical Development of Quantum Theory*, vol. 2 (New York: Springer-Verlag), p. 77.

5. Letter, Born to Bohr, 16 April 1924. Quoted (in German) in Niels Bohr, *Collected Works*, vol. 5, *The Emergence of Quantum Mechanics (Mainly 1924–1926)*, Klaus Stolzenburg, ed. (Amsterdam: North-Holland, 1984), p. 299.

6. Letter, Pauli to Bohr, 11 February 1924. Quoted in Mehra and Rechenberg, vol. 2, p. 133.

7. Heisenberg interview in Archives for the History of Quantum Physics, p. 4.

8. Letter, Heisenberg to Bohr, 21 April 1925. Quoted in Mehra and Rechenberg, vol. 2, p. 209.

9. Letter, Heisenberg to Bohr, 16 May 1925. Quoted in Mehra and Rechenberg, vol. 2, p. 213.

10. Hendrik Kramers, quoted in Mehra and Rechenberg, vol. 2, p. 214.

11. Helge S. Kragh, *Dirac: A Scientific Biography* (Cambridge: Cambridge University Press, 1990), p. 20.

12. J. H. Van Vleck, quoted in Kragh, p. 22.

13. Wolfgang Pauli, *Zeitschrift für Physik 36* (1926), 336–363, p. 357.

14. Letter, Niels Bohr to Wolfgang Pauli, 13 November 1925, quoted in Mehra and Rechenberg, vol. 3, p. 181.

15. Letter, Werner Heisenberg to P. A. M. Dirac, 9 April 1926, quoted in Kragh, p. 23.

16. Max Born, lectures at MIT, 1926. Quoted in Mehra and Rechenberg, vol. 3, p. 171.

8. The Hydrogen Atom

1. Letter, Schrödinger to Einstein, 3 November 1925. Quoted in Walter Moore, *Schrödinger: Life and Thought* (Cambridge: Cambridge University Press, 1989), p. 192.

2. E. Schrödinger, *Annalen der Physik 79*, 361–376 (1926), p. 361. For a translation of this paper, see E. Schrödinger, *Collected Papers on Wave Mechanics* (London: Blackie & Sons Limited, 1928), 1–12, p. 1.

3. Ibid., p. 8.

4. Ibid., p. 9.

5. Letter, Pauli to Jordan, 12 April 1926; quoted in Jagdish Mehra and Helmut Rechenberg, *The Historical Development of Quantum Mechanics*, vol. 5, pt. 2 (New York: Springer-Verlag, 1987), p. 617.

6. Letter, Einstein to Schrödinger, 16 April 1926, quoted in Mehra and Rechenberg, vol. 5, pt. 2, p. 625.

7. Max Born, *Physics in My Generation* (New York: Springer-Verlag, 1969), p. 89.

8. Quoted in Max Jammer, *The Conceptual Development of Quantum Mechanics* (New York: McGraw-Hill, 1966), p. 283.

9. Letter, Schrödinger to Wien, 25 August 1926; quoted in Mehra and Rechenberg, vol. 5, pt. 2, p. 827.

10. Quoted in Abraham Pais, *Inward Bound: Of Matter and Forces in the Physical World* (Oxford: Clarendon Press, 1986), p. 255.

11. Schrödinger, *Collected Papers*, p. 45.

12. Letter, Heisenberg to Pauli, 8 June 1926. Pauli's letter appears in full in B. L. van der Waerden, "From Matrix Mechanics and Wave Me-

chanics to Unified Quantum Mechanics," in *The Physicist's Conception of Nature*, Jagdish Mehra, ed. (Dordrecht: D. Reidel Publishing, 1973), p. 278.

13. Letter, Bohr to Schrödinger, 11 September 1926. Quoted in Mehra and Rechenberg, vol. 2, p. 625.

14. Werner Heisenberg, *Physics and Beyond: Encounters and Conversations* (New York: Harper & Row, 1971), p. 73.

15. Ibid., p. 75.

16. Erwin Schrödinger, *Naturwissenschaften 23*, 807–8812, 823–828, 844–849 (1935).

17. C. Monroe, D. M. Meekhof, B. E. King, and D. J. Wineland, "A 'Schrödinger Cat' Superposition State of an Atom," *Science 272*, 1131–1136 (1996).

9. The Hydrogen Atom and Dirac's Theory of the Electron

1. P. A. M. Dirac, "The Quantum Theory of the Electron, I", *Proceedings of the Royal Society (London) 117*, 610–624 (1928), p. 610.

2. Quoted in Helge S. Kragh, *Dirac: A Scientific Biography* (Cambridge: Cambridge University Press, 1990), p. 57.

3. Quoted in Laurie M. Brown and Helmut Rechenberg, "Dirac and Heisenberg—A Partnership in Science," in *Reminiscences about a Great Physicist: Paul Adrien Maurice Dirac*, Behram N. Kursunoglu and Eugene P. Wigner, eds. (Cambridge: Cambridge University Press, 1987), p. 148.

4. Kragh, p. 65.

5. Ibid., p. 60.

6. Ibid., p. 62.

7. Ibid., p. 63.

8. P. A. M. Dirac, "Quantised Singularities In the Electromagnetic Field," *Proceedings of the Royal Society A133*, 60–72 (1931), p. 61.

9. Kragh, p. 116.

10. Hydrogen Guides Nuclear Physicists

1. Ernest Rutherford and Frederick Soddy, *Philosophical Magazine 4*, 370–390 (1902).

2. Frederick Soddy, "Inter-atomic Charge," *Nature 92*, 399–400 (1913).

3. There is a third isotope of hydrogen called tritium with two neutrons joining the proton in the nucleus. It is radioactive, however, and is

not present in naturally occurring samples of hydrogen. Tritium can be created in the laboratory.

4. Ferdinand G. Brickwedde, "Harold Urey and the Discovery of Deuterium," *Physics Today,* September 1982, 34–39.

5. Ibid.

6. John S. Rigden, *Rabi: Scientist and Citizen* (New York: Basic Books, 1987); republished by Harvard University Press with a new preface (2000), p. 90.

7. For a complete account of the naming of deuterium, see Roger H. Stuewer, "The Naming of the Deuteron," *American Journal of Physics 54,* 206–218 (1986).

11. Hubris Meets Hydrogen

1. Quoted in Abraham Pais, *Inward Bound* (Oxford: Oxford University Press, 1986), p. 208.

2. Quoted in Bretislav Friedrich and Dudley Herschbach, *Daedalus,* Winter 1998, p. 179.

3. John S. Rigden, "Atoms, Molecules and Clusters," *Zeitschrift für Physik D 10,* 119–120 (1988), p. 119.

4. Immanuel Estermann, S. N. Foner, ed., "History of Molecular Beam Research," *American Journal of Physics 43,* 661–671 (1975), p. 670.

12. The Magnetic Resonance Method

1. John S. Rigden, *Rabi: Scientist and Citizen* (New York: Basic Books, 1987); republished with new preface by Harvard University Press (2000), p. 61.

2. Stern, foreword to Ronald G. J. Fraser, *Molecular Rays* (Cambridge: Cambridge University Press, 1931), x.

3. Rigden, p. 104.

4. Ibid.

5. Ibid.

6. Ibid., p. 107.

7. Ibid.

8. Ibid., p. 109.

9. Ibid., p. 96.

10. Ibid., p. 99.

11. Personal conversation, late 1987.

13. New Nuclear Forces Required

1. J. M. B. Kellogg, I. I. Rabi, N. F. Ramsey, Jr., and J. R. Zacharias, "An Electrical Quadrupole Moment of the Deuteron," *Physical Review* 57, 677–695 (1940).

2. Van Vleck to Rabi, 28 November 1938, Rabi's private papers.

3. Van Vleck to Rabi, 5 December 1938, Rabi's private papers.

4. Interview of Hans Bethe by Charles Weiner and Jagdish Mehra, 28–29 October 1966, Center for the History of Physics, American Institute of Physics.

14. Magnetic Resonance in Bulk Matter (NMR)

1. Roger Stuewer, "Bringing the News of Fission to America," *Physics Today*, October 1985, 48–56.

2. James Phinney Baxter, III, *Scientists Against Time* (Boston: Little Brown, 1946), p. 142.

3. Many of the details that follow were drawn from Professor Robert Pound's detailed records. I am grateful to Pound for this information.

4. Edward M. Purcell, AIP Oral History, 1977, p. 27.

5. C. G. Montgomery, R. H. Dicke, and E. M. Purcell, *Principles of Microwave Circuits*, Radiation Laboratory Series vol. 8 (New York: McGraw Hill, 1948).

6. John S. Rigden, "Quantum States and Precession: The Two Discoveries of NMR," *Reviews of Modern Physics 58*, 433–448 (1986).

7. J. H. Van Vleck, "The Absorption of Microwaves by Uncondensed Water Vapor," *Physical Review 71*, 425–433 (1947).

8. E. M. Purcell, H. C. Torrey, and R. V. Pound, "Resonance Absorption by Nuclear Magnetic Moments in a Solid," *Physical Review 69*, 37–38 (1946).

9. F. Bloch, W. W. Hansen, and M. Packard, "Nuclear Induction," *Physical Review 69*, 127, 680 (1946); "The Nuclear Induction Experiment," *70*, 474–485 (1946).

15. Hydrogen's Challenge to Dirac Theory

1. Quoted in Silvan S. Schweber's account of the Shelter Island Conference in his excellent book *QED and the Men Who Made It: Dyson,*

Feynman, Schwinger, and Tomonaga (Princeton: Princeton University Press, 1994), p. 175.

2. Ibid., p. 156.

3. Quoted by K. K. Darrow in ibid.

4. W. V. Houston, "A New Method of Analysis of the Structure of H_α and D_α," *Physical Review 51*, 446–449 (1937); R. C. Williams, "The Fine Structures of H_α and D_α under Varying Discharge Conditions," *Physical Review 54*, 558–567 (1938).

5. Schweber, pp. 218–219.

16. The Hydrogen Atom Portends an Anomaly with the Electron

1. John S. Rigden, *Rabi: Scientist and Citizen* (New York: Basic Books, 1987), p. 177.

2. J. E. Nafe, E. B. Nelson, and I. I. Rabi, "The Hyperfine Structure of Atomic Hydrogen and Deuterium," *Physical Review 71*, 914–915 (1947).

3. Remember, it was Dirac theory that persuaded physicists that they knew the magnetic moment of the proton. Stern's experiment showed that the physicists' assumption was wrong.

4. Rigden, p. 175.

5. Richard Feynman, *QED* (Princeton: Princeton University Press, 1985), p. 118.

17. Hydrogen Maps the Galaxy

1. Quoted in Dieter B. Herrmann, *The History of Astronomy from Herschel to Hertzsprung* (Cambridge: Cambridge University Press, 1973), p. 69.

2. Ibid.

3. J. S. Hey, *The Evolution of Radio Astronomy*, Science History Publications, Neale Watson (New York: Academic Publications, 1973), p. 7.

4. Quoted in W. T. Sullivan III, ed., *The Early Years of Radio Astronomy* (Cambridge: Cambridge University Press, 1984), p. 137.

5. Nicolaas Bloembergen, "Edward M. Purcell (1912–1997)," *Nature 386*, 662 (1997).

6. Unpublished image constructed by Gart Westerhout from Dutch and Australian data.

7. Benjamin K. Malphrus, *The History of Radio Astronomy and the National Radio Astronomy Observatory* (Malabar, Fla.: Krieger Publishing Co., 1996). See p. 26.

8. Ibid., pp. 96–100.

18. The Hydrogen Maser

1. Quoted in "Timekeepers—An Historical Sketch," by H. Alan Lloyd in *The Voices of Time*, J. T. Fraser, ed. (New York: George Braziller, 1966), p. 390.

2. Rabi private collection.

3. J. R. Zacharias, J. G. Yates, and R. D. Haun, "An Atomic Frequency Standard," *Proceedings of the Institute of Radio Engineers 43* (3), 364 (1955).

4. H. M. Goldenberg, D. Kleppner, and N. F. Ramsey, "Atomic Hydrogen Maser," *Physical Review Letters 5*, 361–362 (1960).

5. Norman F. Ramsey, *Spectroscopy with Coherent Radiation: Selected Papers of Norman F. Ramsey with Commentary* (Singapore: World Scientific, 1998), p. 205.

6. R. F. C. Vessot et al., "Test of Relativistic Gravitation with a Space Borne Hydrogen Maser," *Physical Review Letters 26*, 2081–2084 (1980).

19. The Rydberg Constant

1. Quoted in Nadia Robotti, "The Spectrum of ζ Puppis and the Historical Evolution of Empirical Data," *Historical Studies of the Physical Sciences 14*, Part 1, 123–145, p. 125.

2. G. W. Series, "The Rydberg Constant," *Contemporary Physics 14*, 49–68 (1974), p. 51.

3. G. W. Series, *The Spectrum of Atomic Hydrogen* (Oxford: Oxford University Press, 1957).

4. T. W. Hänsch, I. S. Shahin, and A. L. Schawlow, "Optical Resolution of the Lamb Shift in Atomic Hydrogen by Laser Saturation Spectroscopy," *Nature Physical Science 235*, 63–65 (1972).

5. T. W. Hänsch, "Repetitively Pulsed Tunable Dye Laser for High Resolution spectroscopy," *Applied Optics 11*, 895–898 (1972).

6. T. W. Hänsch, M. H. Nayfeh, S. A. Lee, S. M. Curry, and I. S. Shahin, "Precision Measurement of the Rydberg Constant by Laser Saturation Spectroscopy of the Balmer α Line in Hydrogen and Deuterium," *Physical Review Letters 32*, 1336–1340 (1974).

7. C. Wieman and T. W. Hänsch, *Physical Review Letters 36*, 1170–1173 (1976).

8. E. A. Hildum, U. Boesi, D. H. McIntyre, R. G. Beausoleil, and T. W. Hänsch, "Measurement of the 1S–2S Frequency in Atomic Hydrogen," *Physical Review Letters 56*, 576–579 (1986), p. 576.

9. Ibid.

10. T. Andreae, W. König, R. Wynands, D. Leibfried, F. Schmidt-Kaler, D. Meschede, and T. W. Hänsch, "Absolute Frequency Measurement of the Hydrogen 1S–2S Transition and a New Value of the Rydberg Constant," *Physical Review Letters 69*, 1923–1926 (1992).

11. Ibid., p. 1923.

12. Th. Udem, A. Huber, B. Gross, J. Reichert, M. Prevedelli, M. Weitz, and T. W. Hänsch, "Phase-Coherent Measurement of the Hydrogen 1S–2S Transition Frequency with an Optical Frequency Interval Divider Chain," *Physical Review Letters 79*, 2646–2649 (1997).

13. Ibid., p. 2646.

14. A. Huber, Th. Udem, B. Gross, J. Reichert, M. Kourogi, K. Pachucki, M. Weitz, and T. W. Hänsch, "Hydrogen-Deuterium 1S–2S Isotope Shift and the Structure of the Deuteron," *Physical Review Letters 80*, 468–471 (1998).

15. Ibid.

20. The Abundance of Deuterium

1. An English translation of Friedmann's 1922 paper appears in *Cosmological Constants: Papers in Modern Cosmology*, J. Bernstein and G. Feinberg, eds. (New York: Columbia University Press, 1986).

2. Quoted in Helge Kragh, *Cosmology and Controversy* (Princeton: Princeton University Press, 1999), p. 105.

3. R. A. Alpher, H. Bethe, and G. Gamow, "The Origin of the Chemical Elements," *Physical Review 73*, 803–804 (1948).

4. H. Bondi and T. Gold, "The Steady-state Theory of the Expanding Universe," *Monthly Notices of the Royal Astronomical Society 108*, 252–270 (1948) and F. Hoyle, "A New Model for the Expanding Universe," *Monthly Notices of the Royal Astronomical Society 108*, 373–382 (1948).

5. A lovely historical account of cosmology is the recent book by Helge Kragh, *Cosmology and Controversy: The Historical Development*

of Two Theories of the Universe (Princeton: Princeton University Press, 1996).

6. R. A. Alpher and R. C. Hermann, "Evolution of the Universe," *Nature 162*, 774–775 (1948).

7. Quoted in Dennis Overbye, *New York Times*, February 10, 1998.

8. Margaret J. Geller, "Is Cosmology Solved? A Tribute to David N. Schramm," *Astronomical Society of the Pacific 111*, 253 (1999).

9. Overbye, n. 7.

10. Terry P. Walker, Gary Steigman, David N. Schramm, Keith A. Olive, and Ho-Shik Kang, "Primordial Nucleosynthesis Redux," *Astrophysical Journal 376*, 51–69 (1991).

11. Richard I. Epstein, James M. Lattimer, and David N. Schramm, "The Origin of Deuterium," *Nature 263*, 198–202 (1976).

12. Scott Burles, Kenneth M. Nollett, James W. Truan, and Michael S. Turner, "Sharpening the Predictions of Big-Bang Nucleosynthesis," *Physical Review Letters 82*, 4176–4179 (1999), p. 4176.

13. A. Songalla, L. L. Cowle, C. J. Hogan, and M. Rugers, "Deuterium Abundance and Background Radiation Temperature in High-Redshift Primordial Clouds," *Nature 368*, 599–604 (1994).

14. S. Burles and D. Tyler, "The Deuterium Abundance toward Q50 1009+2956," *Astrophysical Journal 507*, 732–744 (1998).

21. Antihydrogen

1. C. S. Wu, E. Ambler, R. W. Hayward, D. D. Hoppes, and R. P. Hudson, "Experimental Test of Parity Conservation in Beta Decay," *Physical Review 105*, 1413–1414 (1957).

2. Richard L. Garwin, Leon M. Lederman, and Marcel Weinrich, "Observation of the Failure of Conservation of Parity and Charge Conjugation in Meson Decays: The Magnetic Moment of the Free Muon," *Physical Review 105*, 1415–1417 (1957).

3. J. H. Christenson, J. W. Cronin, V. L. Fitch, and R. Turlay, "Evidence for 2π Decay of the K_L Meson," *Physical Review Letters 13*, 138–140 (1964).

4. Carl D. Anderson, "The Positive Electron," *Physical Review 43*, 491–494 (1933). The positron originated in Earth's atmosphere.

5. O. Chamberlain, E. Segrè, C. Wiegand, and T. Ypsilantis, "Observations of Antiprotons," *Physical Review 100*, 947–950 (1955).

6. Quoted in "Antihydrogen" by John Eades, Richard Hughes, and Claus Zimmermann, in *Physics World 6*, 44–48, July (1993), p. 48.

7. G. Baur, "Production of Antihydrogen," *Physics Letters B 368*, 251–258 (1996).

8. G. Blanford, D. C. Christian, K. Gollwitzer, M. Mandelkern, C. T. Munger, J. Schultz, and G. Zioulas, "Observation of Antihydrogen," *Physical Review Letters 80*, 3037–3040 (1998).

9. Steven Weinberg, *Dreams of a Final Theory* (New York: Pantheon Books, 1992).

22. The Bose-Einstein Condensate for Hydrogen

1. S. N. Bose, "Plancks Gesetz und Lichtquantenhypothese," *Zeitschrift für Physik 26*, 178–181 (1924).

2. For a complete account of the story see William A. Blanpied, "Satyendranath Bose: Co-Founder of Quantum Statistics," *American Journal of Physics 40*, 1212–1220 (1972).

3. Martin Klein, "Einstein and Wave-Particle Duality," *Natural Philosopher 3*, 1–49 (1963), p. 26; quoted in Blanpied.

4. Albert Einstein, "Quantentheorie des einatomigen idealen Gases," *Berliner Berichte 3*, 3–14 (1925).

5. P. A. M. Dirac, "On the Theory of Quantum Mechanics," *Proceedings of the Royal Society 112*, 661–677 (1926); Enrico Fermi, "Zur Quantelung des idealen einatomigen Gases," *Zeitschrift für Physik 36*, 902–912 (1926).

6. Eric A. Cornell and Carl E. Wieman, "The Bose-Einstein Condensate," *Scientific American* (March) 40–45 (1998).

7. M. H. Anderson, J. R. Ensher, M. R. Matthews, C. E. Wieman, and E. A. Cornell, "Observation of Bose-Einstein Condensation in a Dilute Atomic Vapor," *Science 269*, 198–201 (1995).

8. Daniel Kleppner, "The Yin and Yang of Hydrogen," *Physics Today*, 11–12 (1999).

9. See Issac F. Silvera and Jook Walraven, "The Stabilization of Atomic Hydrogen," *Scientific American* (January) 66–74 (1982).

10. Dale G. Fried, Thomas C. Killian, Lorentz Willmann, David Landhuis, Stephen C. Moss, Daniel Kleppner, and Thomas J. Greytak, "Bose-Einstein Condensation of Atomic Hydrogen," *Physical Review Letters 81*, 3811–3814 (1998).

11. Quoted in *Physics Today*, August 1995, p. 20.

12. M. R. Andrews et al., "Observation of Interference between Two Bose Condensates," *Science 275*, 637–641 (1997).

13. S. Inouye et al., "Phase-Coherent Amplification of Atomic Matter Waves," *Nature 402*, 641–644 (1999).

14. Lene V. Hau et al., "Light Speed Reduction to 17 Meters per Second in an Ultracold Atomic Gas," *Nature 397*, 594–598 (1999).

23. Exotic Hydrogen-like Atoms

1. S. Mohorovicic, *Astronomische Nachrichten 253*, 94 (1934).

2. M. Deutsch, "Evidence for the Formation of Positronium in Gases," *Physical Review 82*, 455–456 (1951).

3. L. W. Alvarez, H. Bradner, F. S. Crawford, Jr., J. A. Crawford, P. Falk-Vairant, M. L. Good, J. D. Gow, A. H. Rosenfeld, F. Solmitz, M. L. Stevenson, H. K. Ticho, and R. D. Triff, "Catalysis of Nuclear Reactions by: Mesons," *Physical Review 105*, 1127–1128 (1956).

4. S. H. Neddermeyer and C. D. Anderson, "Note on the Nature of Cosmic-Ray Particles," *Physical Review 51*, 884–886 (1937); J. C. Street and E. G. Stevenson, "Penetrating Corpuscular Component of the Cosmic Radiation," *Physical Review 51*, 1005 (1937).

5. V. W. Hughes, D. W. McColm, K. Ziock, and R. Prepost, "Formation of Muonium and Observation of Its Larmor Precession," *Physical Review Letters 5*, 63–65 (1960).

6. Vernon W. Hughes, in *Exotic Atoms '79: Fundamental Interactions and Structure of Matter*, Kenneth Crowe, Jean Duclos, Giovanni Fiorentini, and Gabriele Torelli, eds. (New York: Plenum Press, 1980), p. 3.

7. V. Meyer et al., "Measurement of the 1S–2S Energy Interval in Muonium," *Physical Review Letters 84*, 1136–1139 (2000).

8. B. Höglund and Peter G. Mezger, "Hydrogen Emission Line n_{110} 6 n_{109} Detection at 5009 Megahertz in Galactic H II Regions," *Science 150*, 339–348 (1965).

9. Daniel Kleppner, Michael G. Littman, and Myron L. Zimmerman, "Highly Excited Atoms," *Scientific American* (May 1981), pp. 130–149.

10. Michael Nauenberg, Carlos Stroud, and John Yeazell, "The Classical Limit of an Atom," *Scientific American*, June 1994, 44–49.

11. Thomas Stoehlker et al., "1S Lamb Shift in Hydrogenlike Ura-

nium Measured on Cooled, Decelerated Ion Beams," *Physical Review Letters 85*, 3109–3112 (2000).

Epilogue

1. John S. Rigden, *Rabi: Scientist and Citizen*, with a new preface (Cambridge: Harvard University Press, 2000).

Acknowledgments

I. I. Rabi, who died in 1988, figures prominently in this book. He spent much of the 1930s measuring, with ever-increasing accuracy, the magnetic properties of the hydrogen nucleus. While I was writing a biography of Rabi, I got to know him well and as I wrote this book he often seemed to be hovering over my shoulder. Rabi has been an enduring presence and I am grateful for his continuing influence.

Norman F. Ramscy, Jr., himself a student of Rabi, has contributed in a major way to twentieth-century physics through his work on the hydrogen atom as well as other outstanding accomplishments. Ramsey has been a strong supporter of my efforts and has been helpful on many occasions, so I would like to offer a special word of thanks to him.

In many conversations with scientists about this book, my confidence in its basic idea has been reinforced. One such conversation was with Steven Weinberg in Austin, Texas. When our paths crossed in Austin, he asked me, "What are you writing?" I told him about the hydrogen book. After a pregnant pause he said, "That's nice . . . that's nice. Where did you get that idea? Did Rabi give it to you?" Weinberg was referring to my work with Rabi, who was enchanted by hydrogen. Nonetheless, except for the playful suggestion that I was dependent on Rabi for such an idea, I was pleased by Weinberg's response.

Many other physicists (and nonphysicists) have answered my

questions and provided needed information. Among them are Eric J. Chaisson, David Christian, Neil Comins, Eric Cornell, John Eades, Thomas Gallagher, Ted Hänsch (whom I often consulted), Craig Hogan, Wolfgang Ketterle, Tom Kinoshita, Dan Kleppner (whom I consulted again and again), Mel Leon, C. J. Martoff, Bert Mobley, Mary Jo Nye, Sharon O'Dair, William Phillips, Carol Pierman, Ronald Reynolds, Alan Rocke, Philip Schewe, Benjamin Stein, Roger Stuewer, Michael Turner, Fred Ulrich, Gart Westerhout, and Carl Wieman. I have surely missed mentioning some individuals, but to those named and unnamed I express my deep appreciation for their help.

When the book was more an idea than binary code, I received a grant from the Alfred P. Sloan Foundation to support my effort. A major change in my professional life put the manuscript on sustained hold, but no loud noises emanated from the folks at Sloan. I thank the Sloan Foundation for its support and patience.

Michael Fisher and his staff at Harvard University Press have been very generous with their help and advice. I thank them all.

I wrote this book for the general reader—the individual who is interested in the way science progresses and is particularly fascinated by physics. "The general reader" is an abstraction, of course, so I thought of two such "general readers." The first, Eugene Istomin, a world-class concert pianist, reads about physics for pure pleasure. I have enjoyed many conversations with Eugene and I thank him for serving as a shadow reader. Second, there is my son Jonathan, who is a hard-working physician, but reads voraciously and stays informed about advances in physics. He peppers me with questions and sometimes I am hard pressed to provide him with answers. Both Eugene and Jonathan unknowingly guided me in this effort.

Finally, and always, there is my polestar, Diana.

Credits

7.1 Courtesy American Institute of Physics, Emilio Segrè Visual Archives, Segrè Collection.

7.2 Courtesy American Institute of Physics, Emilio Segrè Visual Archives. Photograph by Francis Simon.

8.1 Courtesy American Institute of Physics, Emilio Segrè Visual Archives, *Physics Today* Collection.

9.1 Copyright Cavendish Lab, Cambridge, Great Britain.

12.1 Courtesy American Institute of Physics, Niels Bohr Library.

14.1 Courtesy American Institute of Physics, Meggers Gallery of Nobel Laureates.

14.2 Courtesy Stanford University Archives.

14.3 Courtesy American Institute of Physics, Visual Archives, W. F. Meggers Collection.

14.4 Courtesy Robert V. Pound. Photograph by Mrs. Henry Torrey.

17.2 Courtesy National Radio Astronomy Observatory, Green Bank, West Virginia.

17.3 Courtesy Gart Westerhout.

18 Chapter epigraph courtesy Mrs. Howard Nemerov.

18.1 Courtesy Norman F. Ramsey.

18.2 Courtesy Norman F. Ramsey.

20.1 Courtesy Department of Astronomy and Astrophysics, University of Chicago.

21.1 Courtesy Visual Media Services, Fermilab, Batavia, Illinois.

Index